THE
SOLAR SYSTEM

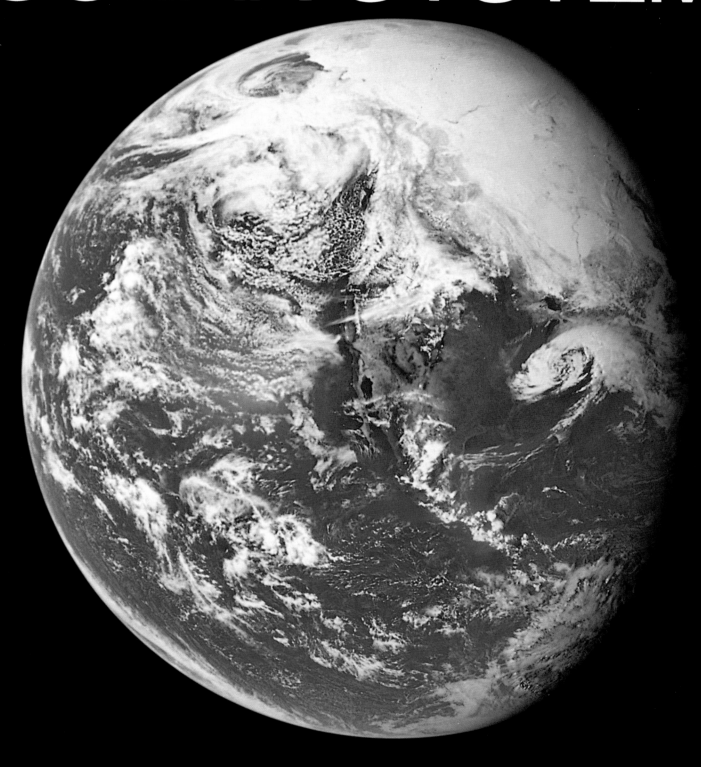

THE SOLAR

GALLERY BOOKS
An imprint of W.H. Smith Publishers Inc.
112 Madison Avenue
New York, New York 10016

SYSTEM

Bill Yenne

Published by Gallery Books
A Division of WH Smith Publishers Inc
112 Madison Avenue
New York, New York 10016

Produced by Brompton Books Corp
15 Sherwood Place
Greenwich, CT 06830

ISBN 0-8317-7888-1

Printed in Hong Kong

10 9 8 7 6 5 4 3 2 1

Edited by Lynne Piade
Designed by Bill Yenne

All photos are courtesy of the National Aeronautics & Space
   Administration (NASA), with the following exceptions:
Max Planck Institut (9, left);
US Naval Observatory (9, right);
Bill Iburg 113; and
Lowell Observatory 115.

Page 1: The planet Earth, as photographed by the crew of
Apollo 16 on their way back from the Moon in April 1972.
Pages 2-3: Seen here are the crescents of Neptune and its
largest moon, Triton, as imaged from a distance of three million miles (4.8 million km) by the Voyager 2 spacecraft in
August 1989.
This page: This reflecting telescope belonged to the English
amateur astronomer William Lassell (1799 to 1880), who discovered Neptune's largest moon, Triton, in 1846.

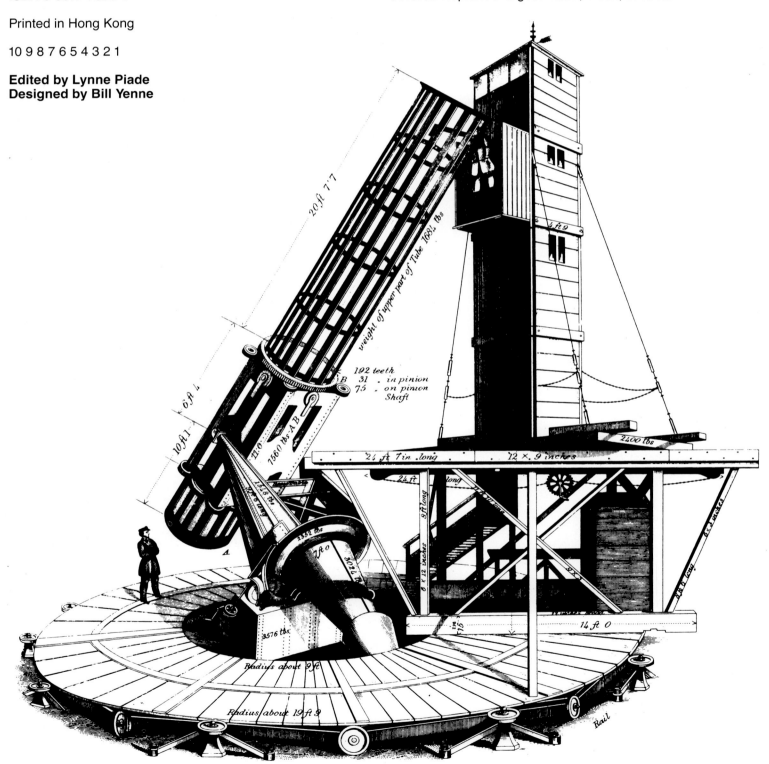

TABLE OF CONTENTS

INTRODUCTION

To the ancients, the planets were the 'wanderers,' the 'stars' that appeared to move most quickly across the heavens. Except for the Sun and Moon, they were also the brightest objects in the sky, and thus were frequently visible in daylight as well as in the dark of night. In the earliest times there were five 'wanderers,' and each was invested with a rich and complex mythology. Many cultures named them for their favorite gods, and some even felt that they *were* the gods.

For most of human history, it was assumed that the stars—the 'wanderers'—the Moon and even the Sun, revolved around the Earth, which was perceived as the center of all creation. In fact, to disbelieve in such a geocentric universe was not just unthinkable, it was heresy!

By the seventeenth century telescopes came into use and mankind began to realize that the wanderers were in fact neither gods, nor even stars, but rather worlds like our own. It was at this time too that we came to realize that our Earth was also a 'wanderer' and that all of us revolved around the Sun. A new concept took hold which was called heliocentrism, meaning that the Sun (which was called Helios by the Greeks) was the centerpiece of a planetary system. The only concession made to geocentrism was that the Moon, and the Moon alone, actually revolved around the Earth.

The heliocentric view tended to lessen the metaphysical importance of the Earth, but it also changed humanity's perception of the other planets. From untouchable gods, they were transformed into mysterious, yet more tangible, places. Once the planets were seen as *worlds* very much like our own, a new mythology arose around the speculation that these worlds were also populated by beings very much like us.

With the advent of telescopes we not only discovered that the Sun, Moon, Earth and five other planets were part of a 'solar system,' but we

At right: The Sun, center of the Solar System, in total eclipse. The 'halo' is the Sun's *corona*, and not seen is its gaseous *photosphere*, just outward of which is the highly active *chromosphere*. The Sun's diameter is 870,331 miles (1.4 million km), and its respective surface and core temperatures are 10,430 degrees and 26.9 million degrees F (5780 degrees and 14.9 million degrees C).

Its equatorial and polar rotation periods are 26.8 and 32 Earth days. The Sun is 93 million miles (150 million km)—one Astronomical Unit (AU)—from Earth.

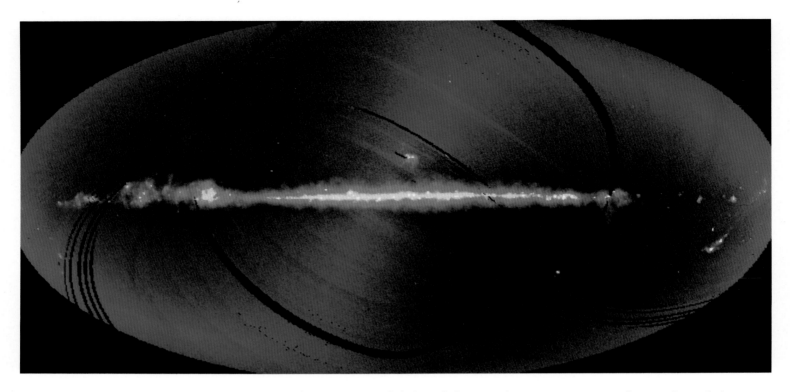

began to locate more objects *within* this solar system than had been known throughout all of the preceding centuries.

In 1610 Galileo Galilei discovered that Jupiter, like Earth, had moons — four of them. Christiaan Huygens identified Saturn's great moon Titan in 1655, and suddenly, the Solar System seemed to be a vastly more complex entity than had been previously been envisioned. A further number of planetary moons were discovered over the next century, and then in 1781, William Herschel confirmed the existence of Uranus, the seventh planet.

Thus, for hundreds of years mankind had believed in the cosmology of five planets. Then, with the advent of telescopes, came the revelation that the Earth was a planet as well. So when Uranus, a heretofore unknown planet, was discovered beyond Saturn, the Solar System expanded to become an almost limitless dimension. After Uranus two more planets were subsequently uncovered in the outer reaches of our Solar System: Neptune in 1846 and Pluto in 1930. Neptune is twice as far from the Sun as Uranus, and Pluto, at its aphelion, is five times farther from the Sun than Saturn. What lies past Pluto's orbit is open to conjecture, but we can no more say that Pluto is the outermost planet of the Solar System today than Galileo should have insisted four centuries ago that Saturn was the outermost planet.

Our Solar System itself originated 4.6 billion years ago when protostellar material, a hot, swirling cloud of mostly pure hydrogen gas (the simplest of elements) gradually collapsed — succumbing to gravity — and

Above far left: An image of the center of our galaxy, the Milky Way, that was obtained by the Infrared Astronomical Satellite (IRAS) in 1983. In this infrared, or 'heat,' photograph, hot galactic material appears as blue or white, while cooler materials appear as red. *At top, above:* Halley's Comet, in a computer reconstruction of an image taken during the comet's celebrated visit to the Earth region of the Solar System in 1910.

Above: A 1986 Halley's Comet closeup—76 years after the 1910 appearance—obtained by the European Space Agency's Giotto probe spacecraft. Note the gas jets that are escaping the nucleus on its Sunward side. Comets are chunks of ice and rock that orbit around the Sun from the outer regions of the Solar System. Some have orbital periods of thousands of years, while others, like Halley, repeat their circuits in mere decades. *Above right:* A time exposure of Comet Ikeya-Seki as it streaked toward the horizon at sunrise in 1965.

cooled. Gravitational contraction heated this protostar and in turn a nuclear fusion reaction was sparked amid the hot, dense gas that condensed at its center, and the Sun was born.

The planets were formed from the remaining disc of material still swirling around the Sun. The four largest planets were, and remain, composed largely of hydrogen as well as helium. As such, they and the Sun are relics of the cloud of protostellar material that existed 4.6 billion years ago. The silicate rock, metals, oxygen, nitrogen, carbon and other materials found in the other planets and moons are probably relics of the impurities that existed in the original cloud. Theoretically, the Solar System extends outward from the Sun to the point (or circular series of points) beyond which the Sun's gravity has no effect. Because theoretical points are hard to measure precisely (and gravity extends indefinitely), we could use the orbital diameter of Pluto, the outermost planet, as the

The Viking 1 and 2 Orbiter/Lander interplanetary probes, which reached Mars in 1976, greatly increased our knowledge of the Red Planet. The Viking Landers (*at right*) were responsible for the first photos taken from the Martian surface. *At top, above:* A Viking 1 image—of its own soil sampler arm against a background of Chryse Planitia, a rolling plain just north of Mars' equator. Martian surface soil consists mainly of oxides of silicon and iron. The thin atmosphere is 95.3 percent carbon dioxide, yet geologic evidence shows that water once flowed freely on Mars—traces of which may remain as permafrost. Mankind's explorations have also touched the outer Solar System.

The intineraries of the Voyager 1 and 2 outer planetary probe spacecraft *(overleaf)* (launched in 1977) included Jupiter, Saturn, Uranus and Neptune.

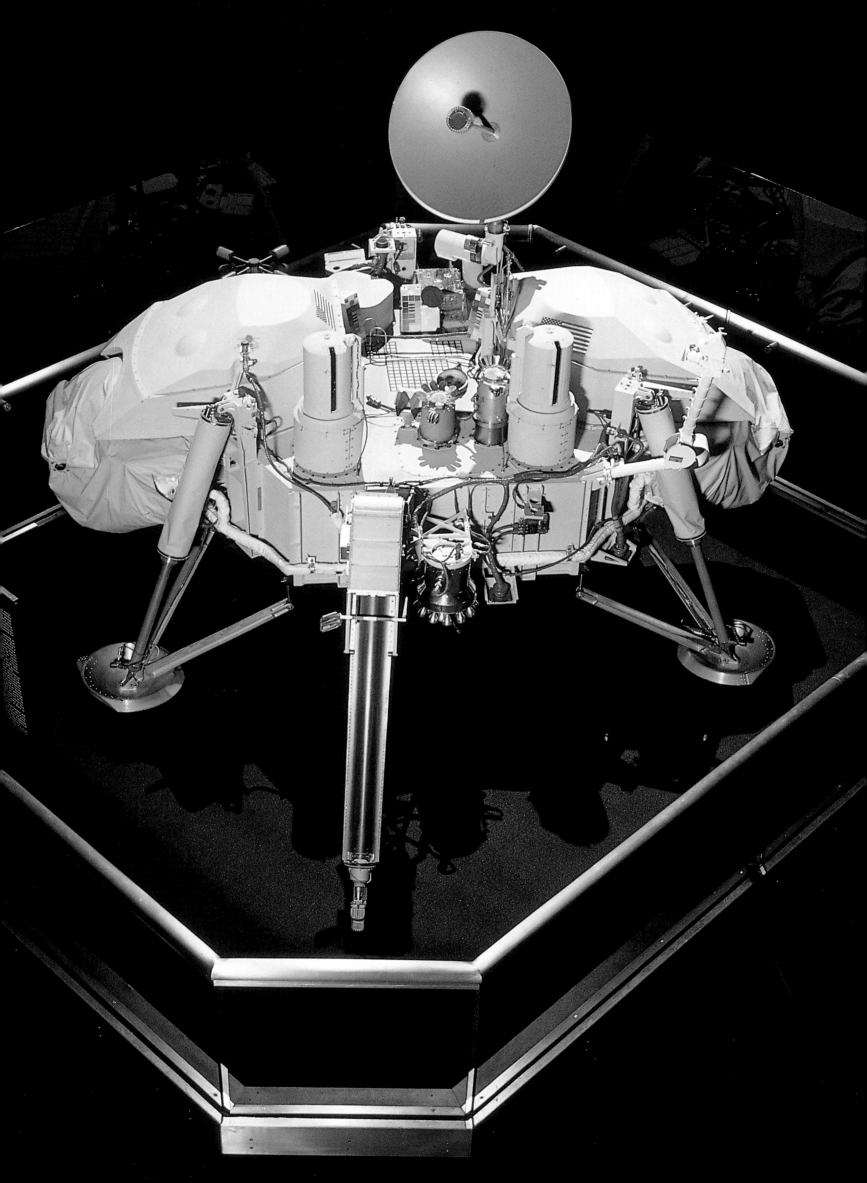

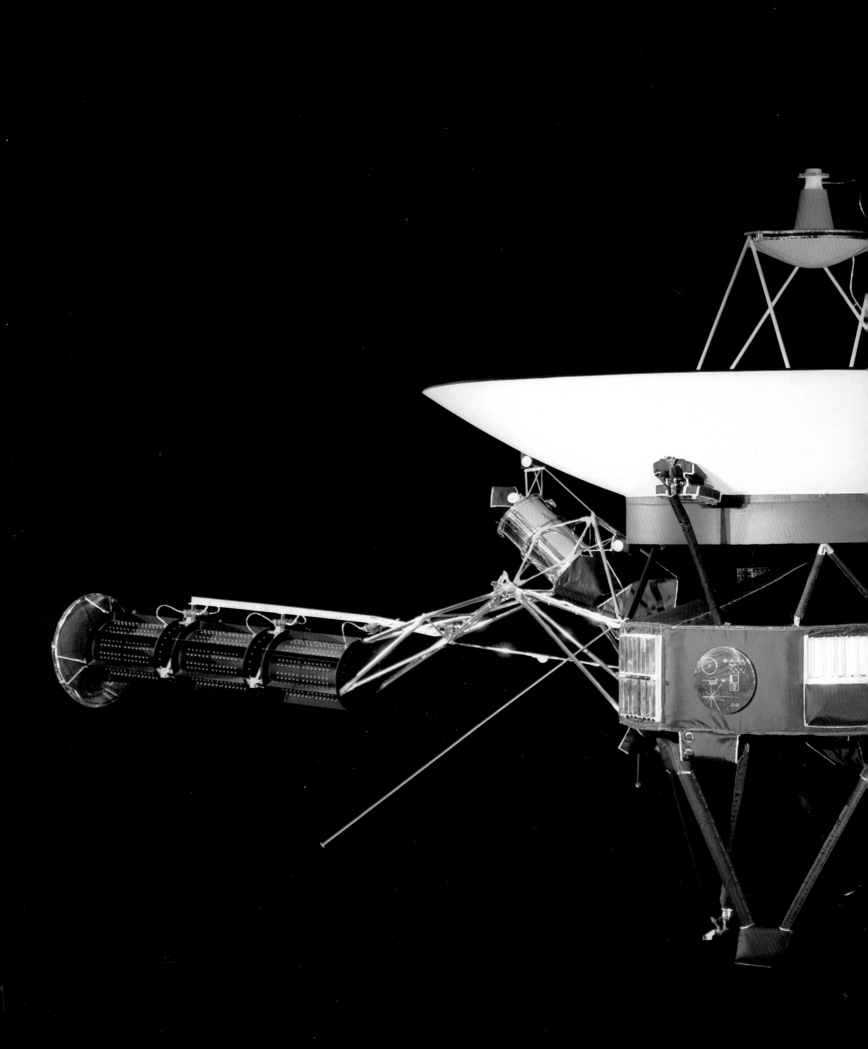

diameter of the Solar System. However, because Pluto's eccentric orbit briefly brings it closer to the Sun than Neptune's more circular orbit, the diameter of what is familiarly known as our Solar System should probably be pegged to the aphelion of Pluto (4.57 billion miles, 7.3 billion km), plus the aphelion of Neptune (2.81 billion miles, 4.5 billion km). Thus the diameter of the Solar System is roughly 7.38 billion miles (11.8 billion km), .00126 light years, or just short of 11 light hours. This formula (or the diameter of Pluto alone) falls short, however, of truly defining the limits of the Solar System. At a distance of 5580 billion (8930 billion km) to 7440 billion miles (11,900 billion km) from the Sun there is a spherical cloud containing comets, with a total mass of seven to eight times that of the Earth. A more massive inner cloud 100 times the mass of the outer cloud stretches from Neptune's orbit to a distance of 930 billion miles (150 billion km). Taking these distances into account would indicate a Solar System diameter on the order of 2.5 light years.

In short, the Solar System can generally be organized into six parts or zones, whose distance from the Sun we are listing here in astronomical Units (AU), which are equal to the distance from the Sun to the Earth, or 93 million miles (150 million km). Moving outward from the Sun they are:

1. The terrestrial, or solid surfaced, planets (Mercury, Venus, Earth and Mars) with their total of only three moons, which span the first 1.6 AU from the Sun.
2. The Asteroid Belt which spans the 3.8 AU distance from the orbit of Mars to the orbit of Jupiter. Most, but not all, known asteroids are to be found within this belt.
3. Beginning 5.4 AU from the Sun, and spanning a distance of 24.8 AU, the third and widest zone includes the four largest planets (Jupiter, Saturn, Uranus and Neptune) along with their 50 moons. These planets are identified as gas giants because of their composition and because they are much larger than any other body in any other zone.
4. The final zone of the familiar Solar System contains the planet Pluto and its single moon, which orbit in an elliptical orbit that ranges between 29.5 AU to 49.2 AU from the Sun.
5. The inner of two clouds of comets that extends from 30 AU to 10,000 AU.
6. The second of two clouds of comets that extends from 60,000 AU to 80,000 AU.

THE PLANETS

	Diameter	Average Distance from the Sun	Number of Known Moons	Closest Visit by a Spacecraft from Earth
Mercury	3031 mi (4878 km)	36 million mi (58 million km)	0	Mariner 10 (1974)
Venus	7521 mi (12,104 km)	67 million mi (108 million km)	0	Venera 11-14 Landers (1978-82)
Earth	7926 mi (12,756 km)	93 million mi (150 million km)	1	—
Mars	4212 mi (6739 km)	141 million mi (228 million km)	2	Viking 1, 2 Landers (1976)
Jupiter*	88,650 mi (142,984 km)	483 million mi (779 million km)	16	Voyager 1, 2 (1979)
Saturn*	74,565 mi (120,000 km)	885 million mi (1.4 billion km)	20 +	Voyager 1, 2 (1980-81)
Uranus*	32,116 mi (51,800 km)	1.7 billion mi (2.9 billion km)	15	Voyager 2 (1986)
Neptune*	30,642 mi (49,424 km)	2.8 billion mi (4.5 billion km)	8	Voyager 2 (1989)
Pluto	1375 mi (2200 km)	3.6 billion mi (5.9 billion km)	1	None (None planned)

The Moons of the Inner Solar System*

	Discovery	Diameter	Distance from Planet
Mercury	None		
Venus	None		
Earth			
Luna (the Moon)	prehistoric	2160 mi (3476 km)	252,698 mi (406,676 km)
Mars			
Phobos	Asaph Hall, 1877	14 mi (23 km)	5760 mi (9270 km)
Deimos	Asaph Hall, 1877	7.5 mi (12 km)	14,540 mi (23,400 km)

*The moons of Jupiter, Saturn, Uranus and Neptune are listed on pages 65, 85, 99 and 109.

The nine known planets can be divided into two distinct groups of four, with the remote, elusive Pluto being in a category of its own. The first group of planets are called the terrestrial (solid-surfaced) planets and they all:

1. Are characterized by solid, rocky surfaces.
2. Are closer than 154 million miles (250 million km) to the Sun.
3. Are *smaller* than 7926 miles (12,756 km) in diameter.
4. Have only three moons among them. (Two of these four planets have *no* moons.)

The second group, the gas giants, all:
1. Are characterized by having a gaseous consistency composed primarily of hydrogen and helium, with a thick, hazy atmosphere composed largely of methane and ammonia.
2. Are *farther* than 500 million miles (800 million km) from the Sun.
3. Are *larger* than 32,000 miles (52,000 km) in diameter.
4. Have more than 50 moons among them, with Saturn alone having over 20.

All of these eight planets revolve around the Sun in a flat ecliptic plane, in orbits which vary by less than three degrees from that plane, except for innermost Mercury, which varies by seven degrees. Pluto, on the other hand, has a wildly different orbital plane, which is inclined by 17 degrees from the plane in which its eight brethren travel around the Sun.

Like our own Earth, each of the planets is a unique, dynamic world unto itself, with distinct features and as yet unsolved mysteries. This book is a capsule view of the best images available of those nine worlds, as seen through the most highly developed telescopes and from the space-craft which have been sent to explore them over the past three decades.

Just as the American Mariner and Soviet Venera inner planetary probes explored Mercury and Venus, the Voyager outer planetary probes explored the Solar System's 'gas giants.' Voyager 1 flew by, collected data on, and took spectacular photos of Jupiter and Saturn, encountering those planets and their moons on 5 March 1979 and 12 November 1980, respectively, before it was purposely ejected from the ecliptic plane.

Voyager 2's itinerary was Jupiter, Saturn, Uranus and Neptune. The space-craft made these encounters on 9 July 1979, 25 August 1981, 24 January 1986 and 25 August 1989, respectively. Voyager 2 is now headed for the edge of the known Solar System, and Earth's Deep Space Network stations will continue auditing its signals well into the twenty-first century. *At right:* An artist's concept of Voyager 2 as it slung around Saturn in 1981, using the giant planet's intense gravitational field as a means to gather momentum for its journey to Uranus and Neptune.

MERCURY

The ancient Chaldeans called it Nego, the planet of warning, but both the Greeks and Romans named this small planet Hermes and Mercury, respectively, for the messenger of their gods because, as observed from Earth, it moves faster than all the other planets, a phenomenon which is the result of its being the closest planet to the Sun. Mercury's year—its period of revolution around the Sun—is just 88 days.

Whether known by the Greek name Hermes or as the wing-footed Mercury, it has been observed and studied by mankind for centuries as the post-sunset 'evening star' or as a predawn 'morning star.' Although both Johann Hieronymus Schroeter—in the early nineteenth century—and Giovanni Schiaparelli—in the early twentieth century—conducted exhaustive telescope studies of Mercury, we did not begin to get a detailed view of the planet's true nature until the American planetary probe Mariner 10 scrutinized it up close in 1974. The Mariner 10 photographs portray Mercury as a barren, rocky world devoid of an atmosphere, whose only distinguishing features are the hundreds of thousands of meteorite impact craters that mar its surface.

At first glance Mercury appears to be very much like Earth's Moon, and indeed, it is very similar in size and texture. However, Mercury has a diameter of 3031 miles (4878 km), half again larger than the Moon, and its cratered face is discernible only to the trained eye. The principle difference between the two is the absence on Mercury of the large, dark patches which are known on our Moon as 'seas.' The 'seas,' which exist only on the side of the Moon closest to the Earth, are composed of deposits of darker basalt rock, which flowed out of the Moon's interior in relatively recent geologic time. Such flows were probably triggered by the earth's gravitational effect on the Moon in much the same way the Moon's gravity generates tides on Earth.

At right: An artist's conception of Mariner 10's Mercury encounter. Mariner 10 began its explorations with Venus on 5 February 1974, and went on to a 29 March encounter with Mercury, making two more Venus flybys on 21 September 1974 and 16 March 1975 (the closest at 168 miles [268 km] altitude). Barren, Sun-blasted and crater-pocked, Mercury is, at 3031 miles (4878 km) in diameter, the Solar System's second smallest planet, and orbits just 36 million miles (58 million km) from the Sun.

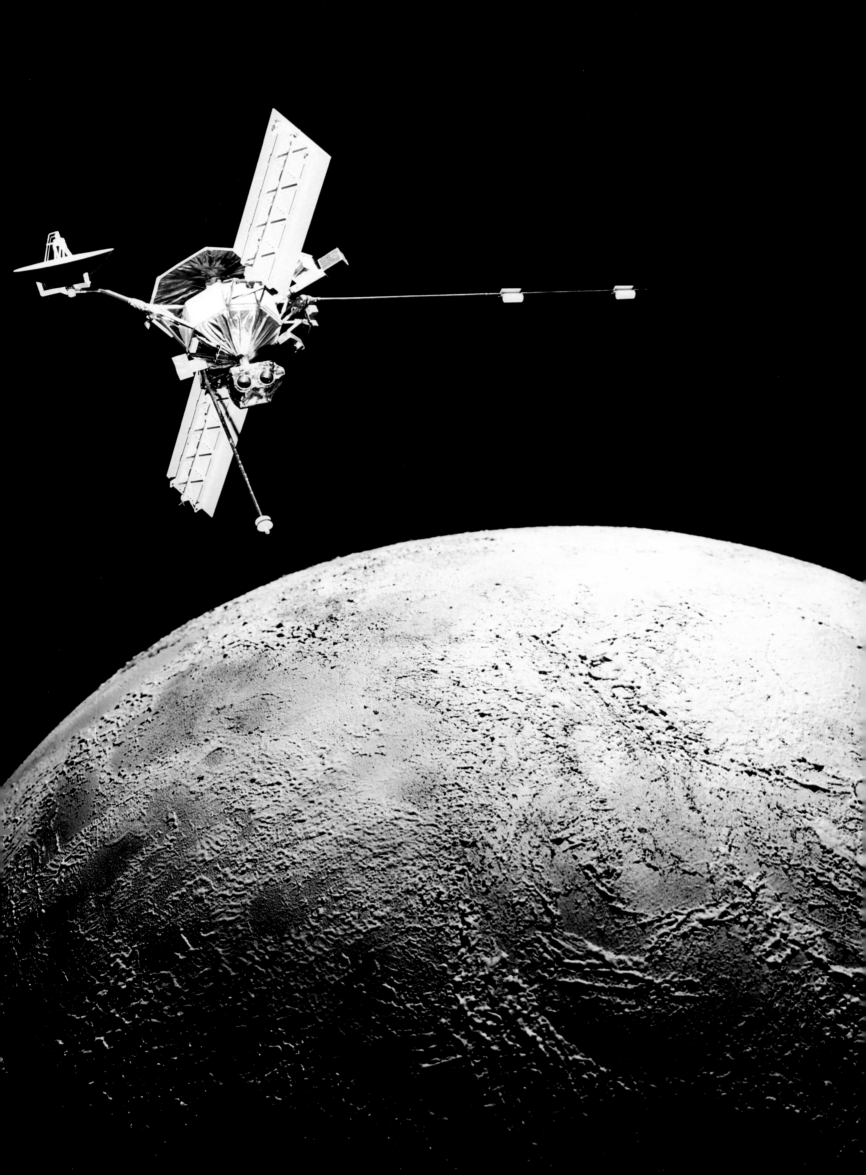

Lacking this type of partnership, Mercury has a much more uniform surface. Its most most prominent feature is the Caloris Basin, a lowland area that is 840 miles (1350 km) across, that is thought to have been caused by a very large, very ancient, meteorite. So extraordinary was this impact that some of the mountain ridges created by it rise as much as 6300 feet (1950 meters) above the floor of the Caloris Basin.

At an average distance of only 36 million miles (58 million km) away from the Sun, Mercury is only a third as far from the Sun as the Earth, and because of this, its surface is a cruel, inhospitable place. In fact, Mercury is close enough to the Sun that lead would flow like water! During the course of the Mercurian day, which is equivalent to 59 Earth days, temperatures on the side facing the Sun can reach 620 degrees F (326 degrees C). Even in the shade, temperatures seldom cool below 200 degrees F (95 degrees C).

During the long Mercurian night, however, the surface chills to colder than −300 degrees F (−150 degrees C). The expansion and contraction of rocks due to this thousand degree fluctuation in temperature no doubt plays an important role in eroding and changing the planet's surface features.

At top, above: A Mariner 10 photomosaic of Mercury's south polar region, from a distance of 40,000 miles (64,000 km) on 21 September 1974 (see caption, page 18). Another photomosaic from this same Mariner 10 Mercury encounter is shown *at right*—the south polar region from a distance of 31,000 miles (49,600 km). Note the bright rays radiating from fresh impact craters: the pole is mid-center of the terminator at the bottom. The large craters in this photo average 43 miles (69 km) across.

VENUS

The Chaldeans called her Ishtar, the Sumerian virgin mother, the Lady of Heaven. To the Greeks she was Aphrodite, goddess of beauty, a sentiment echoed by the Romans, who gave her the name Venus, after their goddess of love. Today, Venus is still an object of beauty, the brilliant evening or morning 'star,' the brightest solar body visible from Earth except for the Sun and Moon.

Located 67 million miles (108 million km) from the Sun, this luminous object is the second of the terrestrial, or solid-surfaced, planets. With a diameter of 7251 miles (12,104 km), Venus is almost exactly the same size as Earth, and, like our own planet, has a solid surface and an atmosphere. Beyond these characteristics, however, the similarities cease. The Venusian atmosphere contains a thick cloud cover which shrouds the entire planet *all* the time, while the Earth's cloud cover obscures no more than half its surface at any given moment. This means that the surface of Venus is not detectable from Earth, and that direct sunlight *never* penetrates the clouds and shines on the surface. Though seventeenth and eighteenth century astronomers, who were unaware of this cloud condition, attempted to 'map' Venus, the 100 percent cloud cover was confirmed as a reality by the twentieth century.

The atmosphere of Venus is almost completely (96 percent) composed of carbon dioxide, although the clouds contain both water vapor and sulfur dioxide, which, when mixed, form sulfuric acid rain squalls inside the clouds. Massive electrical storms are also common, and their lightning, long observed from Earth as 'ashen light,' once was thought to be the lights of Venusian cities.

Since the American Mariner 2 spacecraft, which first flew near Venus in December 1962, a dozen such unmanned planetary probes, launched from both the United States and the Soviet Union, have begun to unlock many of this planet's secrets. For example, no one had any idea of the

Venus' thick carbon dioxide atmosphere is continually roiled by lightning storms and sulfuric acid squalls. The divergence of high-altitude prevailing winds at the planet's equator produces Venus' distinctive 'Y Feature,' which is plainly evident in the Pioneer Venus photograph *at right*, taken from a distance of 29,700 miles (47,500 km) on 10 January 1979. The average atmospheric pressure of Venus is 100 times that found on Earth.

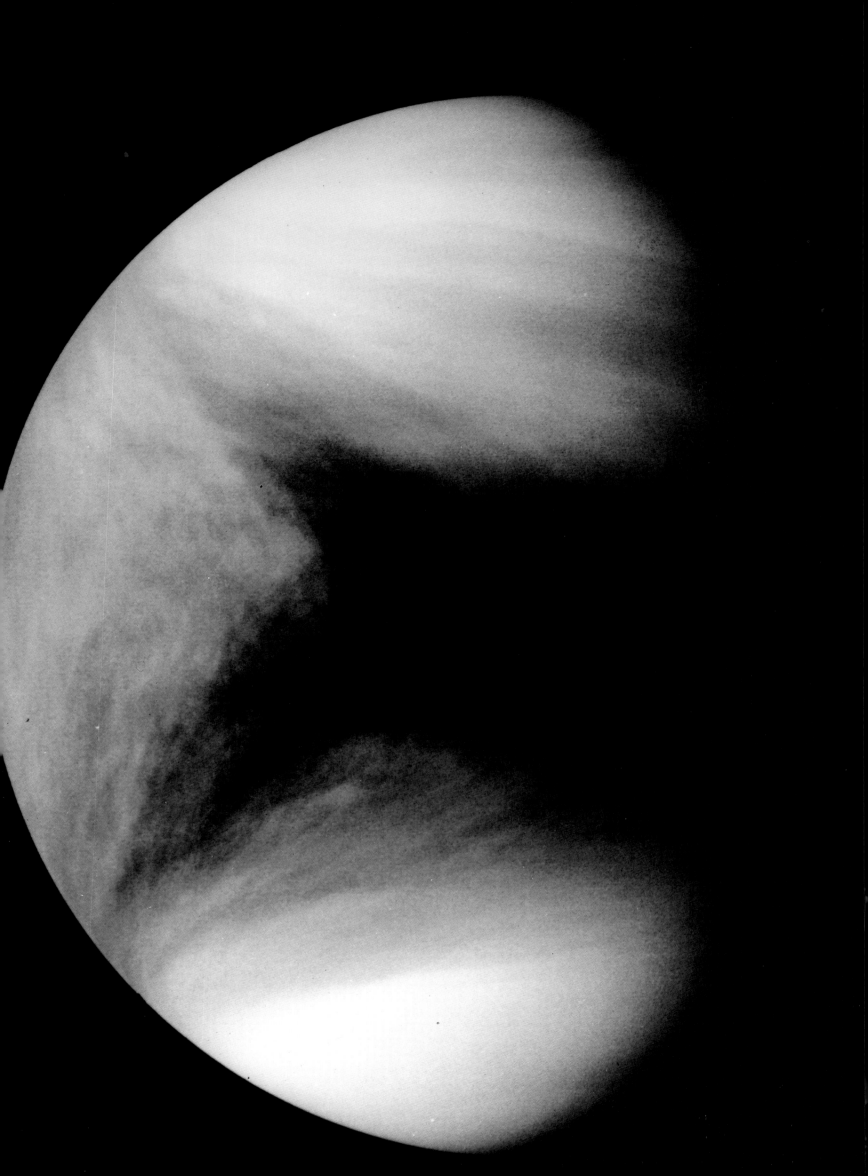

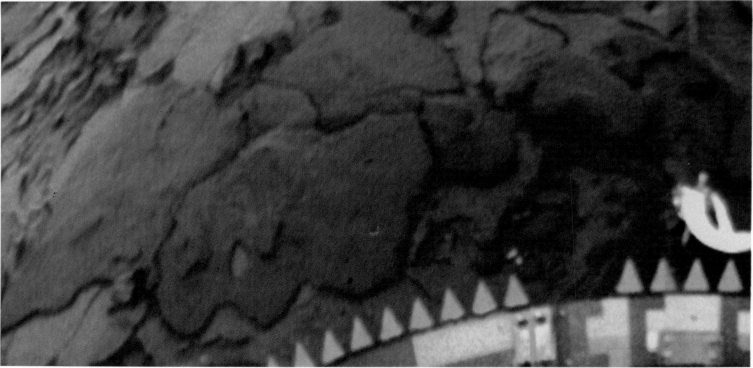

The Soviet Venera 11/12 and 13/14 spacecraft dispensed landers to the Venusian surface. Venera 11/12 survived to transmit data for 95 and 110 minutes, respectively, in the intense pressure and 900-degree F (482 degrees C) heat on the surface, but failed to return planned television pictures. *At top, left and right*, we see the first color pictures of Venus' surface, taken by Venera 13 in 1982. These 'fisheye lens' photos (like the two Venera 14 photos *above*) have a yellow-orange coloration due to sunlight filtering through sulfuric acid

droplets in the cloud cover. The surface air of Venus is relatively calm and clear, but 30 miles (48 km) up the cloud cover begins, continuing for another 15 miles (24 km) in altitude.

This global overcast is pushed by winds of up to 200 mph (320 kph), circulating around the planet once every four Earth days. By contrast, the Venusian day is 243 Earth days long. The clouds produce a greenhouse effect, making the heat a global constant.

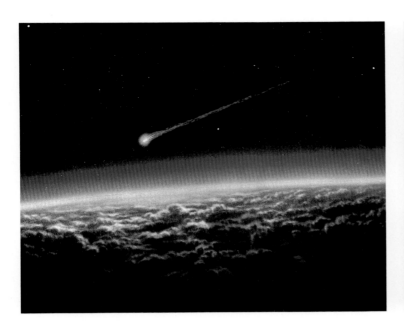

nature of Venus' surface features until 1978, when the American Pioneer Venus spacecraft used radar to 'look' beneath the clouds. The picture painted by the data returned from such spacecraft is not an inviting one. Although some romantics once envisioned lush, steamy jungles beneath Venus' clouds—benefiting from the carbon dioxide and water known to be present—in actual fact, we now know that the surface temperature on Venus, which is kept constant around its globe by the presence of the cloud cover, is a suffocating 900 degrees F (482 degrees C)—far too hot for water to exist in liquid form. Indeed, the view from the surface is that of endless, windy, dusty plains bathed in the smoggy, orange glow of the Venusian day, which lasts the equivalent of 243 Earth days. In fact the Venusian day—the planet's rotational period—is longer than its year—its revolution around the Sun—which is just 225 Earth days.

Even if one could survive the withering heat and depressing haze long enough to inspect the planet from its surface, the atmospheric pressure there would render an effect like that of being crushed by a 20-ton weight. This atmospheric pressure, which is roughly 100 times that encountered on Earth's surface, has made it very difficult to conduct successful spacecraft landings. In fact the only spacecraft to ever reach the surface have been destroyed within minutes. Four Soviet Venera spacecraft, equipped with cameras, landed on the surface between 1978 and 1982, but were only able to squeeze off two pictures apiece before being crushed.

Thanks to the Pioneer Venus project of 1978 and the Magellan project launched by the United States in 1989, we do have something of an idea

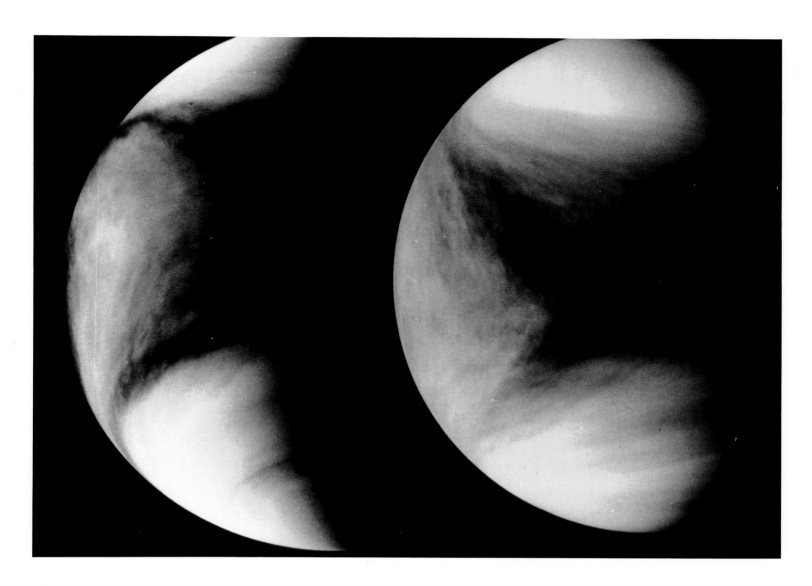

of what the major surface features look like. In the northern hemisphere there is a vast upland region, which has been dubbed Ishtar Terra after the ancient Sumerian goddess of love and fertility. The highest point in this vast, mountainous country is in the Maxwell Mountains, that rise to more than 35,000 feet (107,000 meters) above the planet's mean radius, which is about 20 percent higher than the Earth's Himalayas rise above sea level.

In the southern hemisphere there is another great upland area, which is named Aphrodite Terra after the Greek counterpart of the Roman Venus. Measuring 7000 miles (11,200 km) across, Aphrodite Terra is the planet's largest feature and it also contains the deepest canyon on Venus—the Diana Chasma—whose floor is 9500 feet (2900 meters) below the mean radius of the planet.

At top: Pioneer images, evidencing Y Feature fluctuation (see caption, page 22). *Above far left:* A conception of a probe from the Pioneer Venus Multiprobe entering the Venusian atmosphere in 1978. *Above left:* A conception of a Pioneer probe on Venus' surface, 60 percent of which lies within 1600 feet (500 meters) of Venus' mean radius, in contrast to such features as the 35,000-foot (10,700-meter) Maxwell Mountains.

EARTH

To a space traveler, she looks like a vivid blue marble with vast seas stretching over three-quarters of her surface. Known to her human inhabitants as Mother Earth, this planet's oceans are thought to be the point of origin of life on the only planet in the Solar System known to support life.

Located 93 million miles (150 million km) from the Sun, the Earth is the third of the closely spaced, terrestrial planets. With a diameter of 7926 miles (12,756 km), it is also the largest of the four terrestrial planets.

Composed primarily of nitrogen (76 percent) and oxygen (21 percent), Earth's atmosphere is much *less* dense than that of Venus, yet much *more* dense than that of Mars. Earth's clouds, which are chiefly water vapor, are spread over roughly half the planet in constantly changing patterns. These clouds exist in the seven-mile-deep troposphere, which is the lowest, densest layer of the Earth's atmosphere. Above the troposphere is another layer of atmosphere that is approximately 25 miles (40 km) thick called the stratosphere. It is the ozone present in the stratosphere that prevents solar ultraviolet radiation from reaching the Earth. Above this layer is the mesosphere, which is 20 miles (32 km) thick, and above that the ionosphere, which extends upward for another 100 miles (160 km), to a point about 150 miles (240 km) above the surface.

Manned spacecraft, including the American Space Shuttle and the Soviet space stations, typically operate at altitudes of 250 to 300 miles (400 to 480 km), and thus are well above most of the Earth's atmosphere. Viewed from this altitude, the denser troposphere and stratosphere appear as a narrow blue band on the Earth's horizon, while the other layers are too insubstantial and faint to be visible at all.

Beneath the atmosphere is a world whose environment has proven ideal for the development of a myriad of life forms. Although temperatures range from 136 degrees F (60 degrees C) in the equatorial deserts

At right: A view of Earth, taken from the Apollo 13 spacecraft in 1970. The largest of the terrestrial planets, with a diameter of 7926 miles (12,756 km), it is also the only planet in the Solar System known to support life. Its atmosphere is much thinner than that of Venus, yet is denser than that of Mars. Shown here are Earth's oceans, and her ever-changing clouds—born aloft in the mostly nitrogen-oxygen atmosphere.

to −130 degrees F (−100 degrees C) at the polar ice caps, the global mean temperature is about 60 degrees F (15 degrees C).

The Earth rotates on its axis every 24 hours, which provides a regular pattern of cooling and solar warming that helps to moderate the overall global surface temperature. The water in oceans and seas, that encompass three-quarters of the total surface area, is continuously cycled into the atmosphere by evaporation and back onto the Earth in the form of rain or snow. Water exists in a liquid state over most of the surface a majority of the time, and in the form of ice and snow seasonally, except in the highest mountain areas and in the north and south polar regions— areas which receive the least amount of sunlight.

The solid crust of the Earth is only about 25 miles (40 km) thick and is composed largely of silicate rock. Beneath this surface is a mantle, which

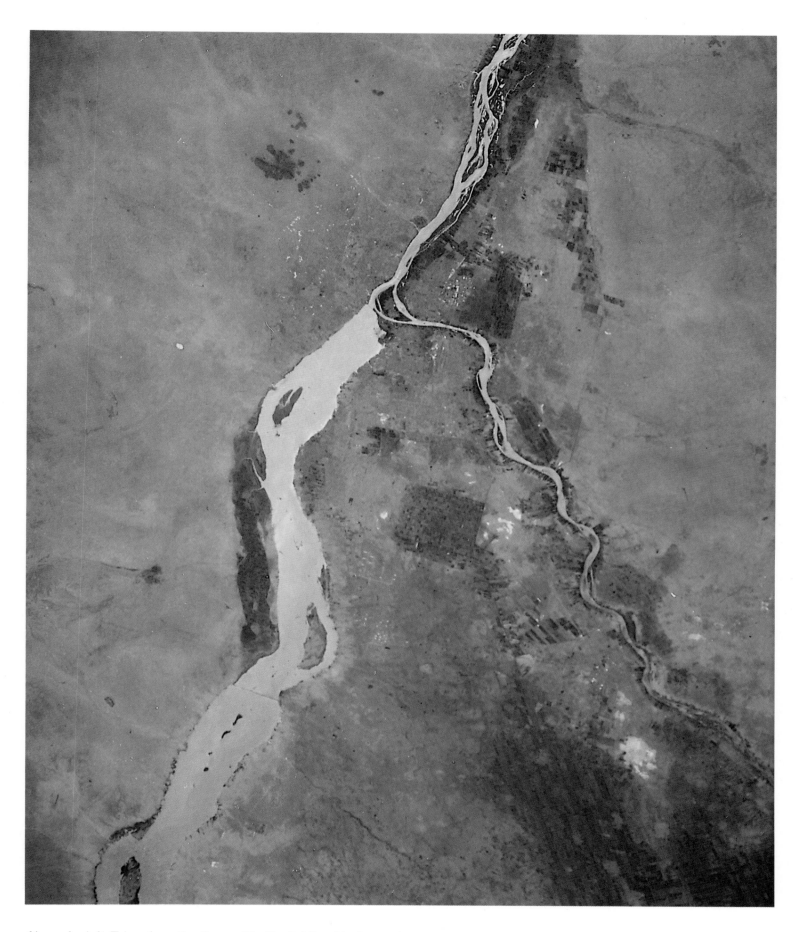

Above far left: Taken from the Space Shuttle Orbiter *Challenger* in 1983, this photo of cumulonimbus cloud formations over Zaire (situated on the equator in central Africa) shows the action of updrafts and water vapor in Earth's atmosphere. *Above:* This Space Shuttle *Discovery* photo shows northern Africa's White and Blue Nile rivers, which here evidence recent flooding (brown areas). Note the White Nile Dam, at photo left.

consists of semi-molten, silicate rock 1800 miles (2900 km) thick. The Earth's core is thought to be a sphere of molten iron (with some nickel), roughly 4200 miles in diameter.

So thin is the Earth's crust that it actually 'floats' on the mantle in the form of semidetached plates. Along the lines— or rifts—where these plates adjoin, the crust is extremely fragile, and it is here that earthquakes occur. Occasionally, too, the molten rock from the mantle will break through the crust along these rifts in the form of volcanic eruptions, resulting in lava floes. These spectacular displays are a constant reminder that the Earth is second only to Jupiter's moon Io as the most geologically active body in the Solar System.

Above: A Landsat 4 satellite false-color image of Death Valley, in southwestern North America. Surrounded by the Panamint, Grapevine, Funeral, Black, Spring, Granite and Avawatz Mountains, Death Valley temperatures have gone as high as 134 degrees F (57 degrees C). *Above far left:* A Space Shuttle view of the Bahamas: note how the light blue waters of the Bahama Bank contrast with the darker, deep ocean waters.

The manned spacecraft Gemini 10 obtained the *above* photograph of atmospheric vortices over the Earth's Arctic Ocean. Plainly visible here is the counterclockwise spiralling motion imparted to the northern hemispheric winds by the Earth's rotation. In contrast to this, analogous motion in the southern hemisphere is clockwise. Not only wind and cloud masses are affected thus, the same phenomenon is observable in whirlpools of all sizes—be they the products of draining sinks, or Norway's Lofoten Islands *Maelstrom*.

On the subject of rotational movement, the Earth rotates on its axis once every 24 hours, and makes one complete circuit around the Sun every 365.25 days. The Earth's inclination to its axis, 23.4 degrees, produces the Earth's seasons, in which the northern and southern hemispheres are alternately closest to the Sun. Only during the vernal and autumnal equinoxes—or twice each year—does the Sun shine directly on the Earth's equator. Throughout Earth's annual seasonal cycle, Earth's equator receives the most sunlight, while the poles receive the least.

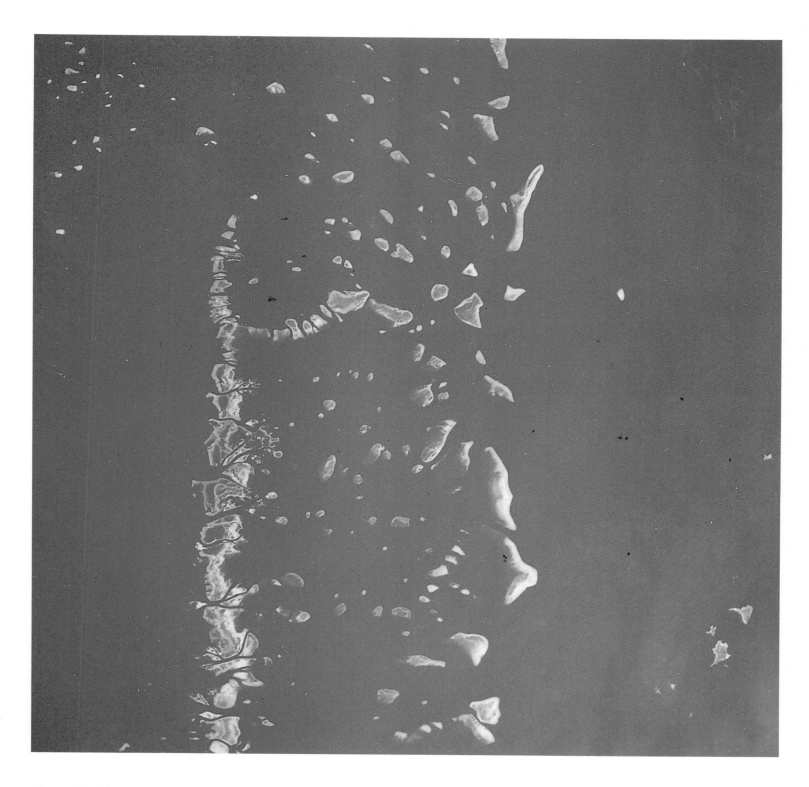

Above: Earth's most extensive ocean reef system, the Great Barrier Reef—off northeastern Australia, near the juncture of the Coral Sea and the Pacific Ocean. Such reefs amply evidence the fecundity of Earth's oceans, and teem with life forms ranging from animals smaller than the tiny creatures whose skeletons formed this reef, to the 35-foot (11-meter) great white sharks (and in the open ocean, the 100-foot [30.4-meter] blue whales).

Earth's continents—(largest to smallest) Eurasia, Africa, North America, South America, Antarctica and Australia—comprise less than a third of the Earth's surface area, while its oceans and seas comprise 70 percent of Earth's surface area. There are four major oceans on Earth—(largest to smallest) the Pacific, the Atlantic, the Indian and the Arctic. The Pacific is 64 million square miles (165.3 million km) in area, and the Arctic is 5.5 million square miles (14.2 million square km) in area.

In contrast, Eurasia is 21.3 million square miles (55 million square km) in area, and Australia, three million square miles (7.8 million square km).

Above: This photo, obtained by a Gemini manned spacecraft in 1966, shows a region where the continental plates of Eurasia and Africa intersect. The bodies of water shown here are, left to right, the Mediterranean Sea, the Gulf of Suez (and the Suez Canal), the Dead Sea, the Gulf of Aqaba, and the Red Sea—from which both of the above gulfs proceed. We are looking north from above the African Continent.

At the top of this view is the lower perimeter of the Eurasian Continent. Countries seen here are Egypt, Jordan, Saudi Arabia, Lebanon, Syria, Iraq, Turkey and Israel.

Temperatures in this region range from approximately 40 degrees F (four degrees C) in the winter months to perhaps 90 degrees F (32 degress C) in summer. It is an arid region, marked by contrasting fertile and barren land. Pointing up this contrast is the photograph *at above right*, taken from a Space Shuttle in 1989. Shown here is the coast of Oman, which would be off-camera to the far right in the previous photo. (Oman lies east, across Saudi Arabia, from the Red Sea.) The dark green of vegetation complements the sandy

desert that dominates the region in this view. Earth's 120-mile-deep (192 km) atmosphere contains the means by which rainfall is distributed to various points on the surface. The lowest layer, the *troposphere*, rises from the surface to a height of seven miles (11 km), where begins the *stratosphere* (or *ozonosphere*), which extends upward to a height of 30 miles (48 km). Above this lies the *mesosphere* (30 to 50 miles) (48 to 80 km) and the *ionosphere* (50 to 150 miles) (80 to 240 km).

Weather, as it is known on Earth, occurs in the troposphere. While the effects of wind and water and the workings of plate tectonics are the basic mechanical processes by which natural change occurs on Earth, the planet's extraordinary plenitude of life forms contribute greatly toward altering or maintaining the environment on Earth.

There are approximately one million animal species (including birds, fish and insects) and 300,000 plant species on Earth. One life form has far greater impact on Earth's environment than the others—mankind, whose influence is felt everywhere on this semi-translucent, blue sphere.

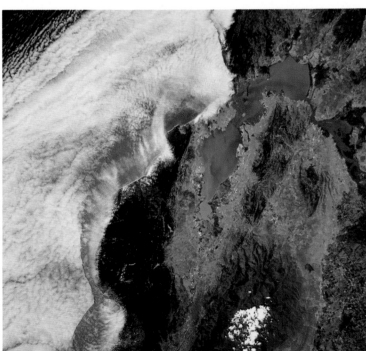

Mankind's desire for knowledge has led to intensive scrutiny via spacecraft of Earth's atmosphere. Most Earth-orbital spacecraft maintain a 250-mile (400-km) altitude above the Earth, beyond the ionosphere.

At top left is an image of a cyclonic storm system developing 1200 miles (1920 km) north of Hawaii, obtained by the manned Apollo 9 spacecraft on 7 March 1969. See also the discussion of atmospheric vortices in the caption on page 34. *At top, above:* An Apollo 7 photo of Hurricane Gladys, southwest of Florida on 18 October 1968. The photo was taken from an altitude of 97 nautical miles (180 km).

Above: An intense rainstorm over the Amazon Basin in Brazil, as imaged from Earth orbit by Apollo 9 in 1969. The Amazon Basin receives some of the Earth's heaviest precipitation.

Above left: Coastal clouds near San Francisco Bay, as imaged by Skylab 2, on 27 June 1973. *At right:* In a September 1966 photo taken by the crew of the manned spacecraft Gemini 11, an Agena test vehicle floats above the lively, and *life-nurturing*, atmosphere of planet Earth.

In contrast to the Earth, and to Venus, Earth's Moon has no atmosphere, though trace amounts of hydrogen and helium (plus hints of argon and neon) were detected escaping from its surface by the Apollo 17 Lunar Lander crew in 1972. The Moon's low mass (3.34×10^{22} kg or .012 that of the Earth) and consequent low gravitational pull account for its inability to entrap the gases and thus establish an atmosphere.

The Moon's diameter is 2160 miles (3476 km), and it revolves around the Earth at a distance of 252,698 miles (406,676 km). *At top, above:* The Moon as imaged by the manned spacecraft Apollo 10, in May 1969. The Earth's only natural satellite, the Moon has captured mankind's imagination for centuries. Linked in mythology to Luna, the Roman goddess of the hunt, the Moon is also that 'lesser light' spoken of in the Bible.

The Moon is illuminated chiefly by the Sun, but during the Moon's crescent phase, Earthshine can be seen faintly illuminating the Moon's disc in contrast to the bright, sunlit crescent portion that indicates that phase. The Moon's rotation is such that the same side always faces the Earth, and its surface is

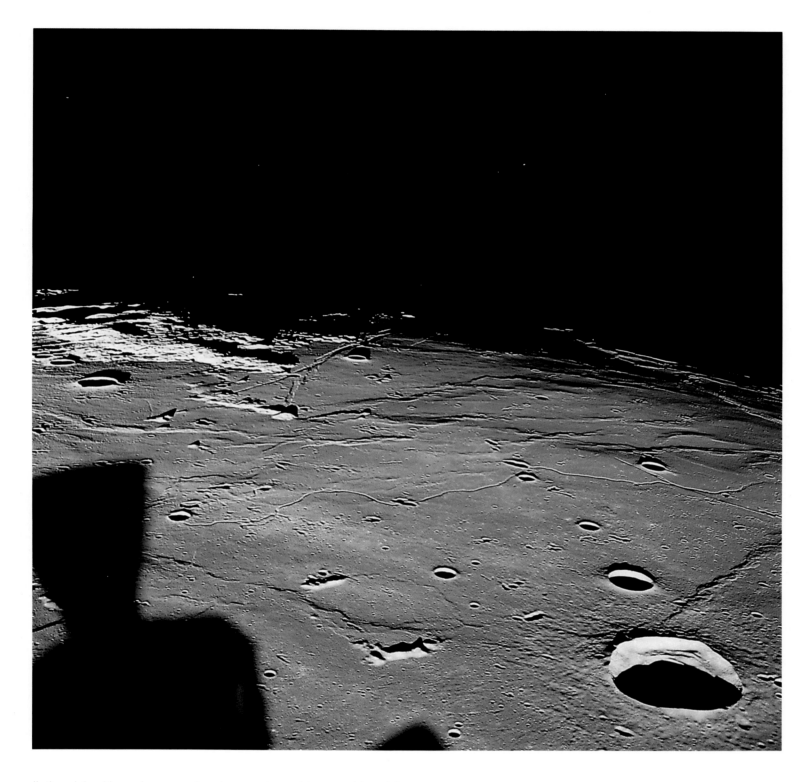

distinguished by volcano and meteor craters, ridges and 'maria'—or seas.
The maria are broad, relatively smooth areas that are solidified, and very
ancient, volcanic lava flows. Most of the maria are grouped on the Earthward side of the Moon, and hence were probably stimulated by Earth's gravitational pull.

Long the subject of much speculation, the Moon's far or 'dark' side is not
really dark at all, but receives as much sunlight—but no Earthshine—as the
near, or 'bright' side. *Above far left:* The large central crater here is Crater Kep-
ler, in the northern hemisphere of the Moon's near side. This image was
obtained by the manned lunar landing mission Apollo 12, in November 1969.
Above: A view of the southwestern sector of the Sea of Tranquility (near side)
evidences a large crater, Maskelyne, and a ridge, Hypatia Rille. The photo was
taken by Apollo 11, during mankind's first lunar landing, on 20 July 1969.

Above left: The Moon is also rife with such mysterious and thought-
provoking formations as this—Davey Crater Chain, as imaged by Apollo 14, in
January 1971.

MARS

To early men, this luminous red mass in the night sky suggested a distant, bloody battlefield. The Chaldeans heralded it as Nergal, the raging king, or the furious one. The Greeks and Romans named it Aries and Mars for their respective gods of war. Astrologers have always attributed malevolent or warlike characteristics to it, although it was long ago understood that Mars' ruddy complexion was due to the strong presence of iron oxide (rust) in its soil, rather than to blood.

At the end of the nineteenth century Giovanni Schiaparelli in Italy and Percival Lowell in the United States studied and mapped what they believed were 'canals' constructed by the Martians to bring water from the ice caps to irrigate fields in the warmer equatorial zones. Even into the twentieth century it was a common and widely-held belief that intelligent beings—much like Earth people—lived on Mars. It seemed that, the more people learned about Mars and uncovered similarities between it and Earth, the more they tended to believe in the existence of Martian people.

Indeed, Mars is very similar to the Earth. Although its diameter of 4212 miles (6739 km) is much smaller than that of the Earth and Venus, Mars resembles Earth in several ways. Its polar ice caps, for example, appear as near twins of those on Earth. The Martian year is 687 Earth days, but the Martian day, at 24.6 hours, is amazingly close to that of Earth's day. Like Venus, Mars has an atmosphere consisting of more than 95 percent carbon dioxide, but oxygen and water vapor are also present, and the Martian equatorial zone can reach temperatures of 80 degrees F (27 degrees C). Such facts as these seemed to point toward the possibility of life on Mars—that is, until 1965 when the American Mariner 4 spacecraft transmitted the first close-up pictures of the planet.

The pictures taken by this and three subsequent Mariners depicted a

At right: A computer-enhanced Viking Orbiter 2 image of Mars' crescent. Shown (photo top to bottom) are the large, frosty Argyre Planitia crater basin; the huge Valles Marineris rift canyon; and Ascraeus Mons, a giant Martian volcano with ice cloud plumes on its western flank. Like Earth, Mars has seasons, including a summer—and a winter that frosts Mars' temperate zones. A Martian year lasts 23 Earth months.

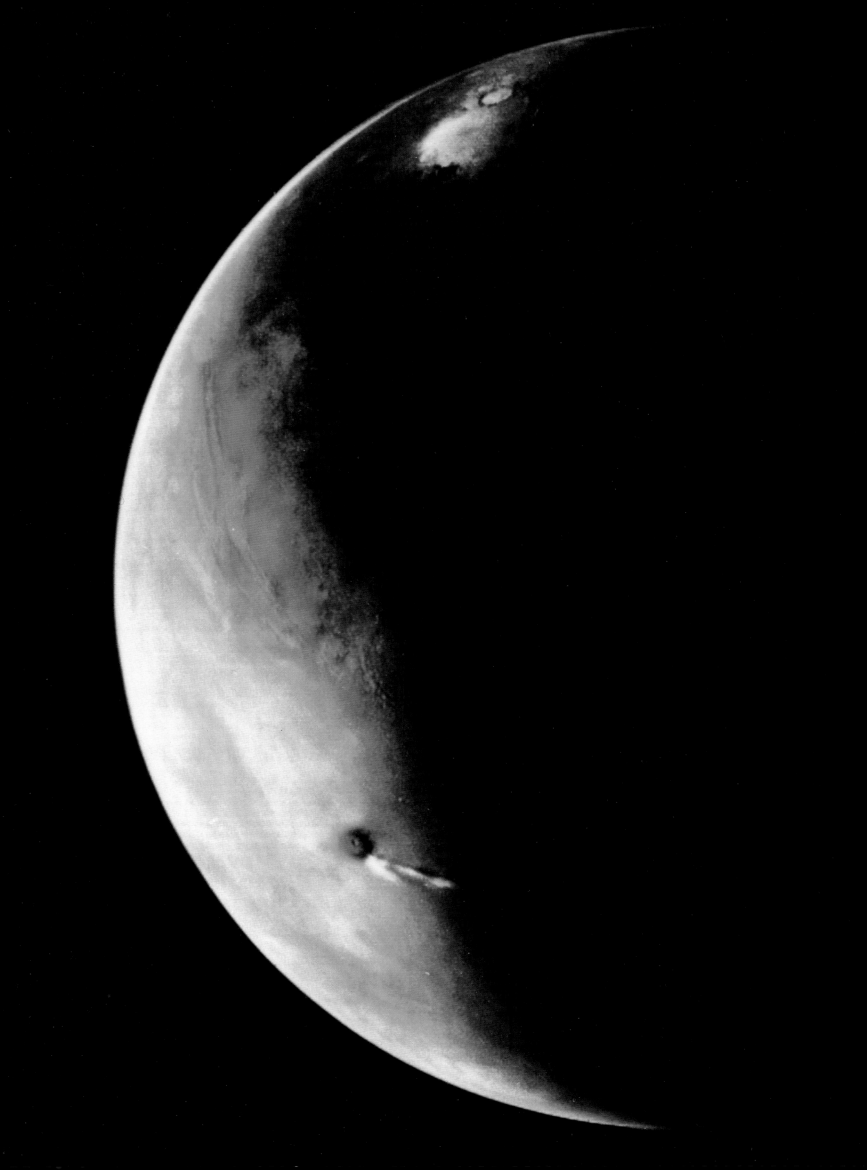

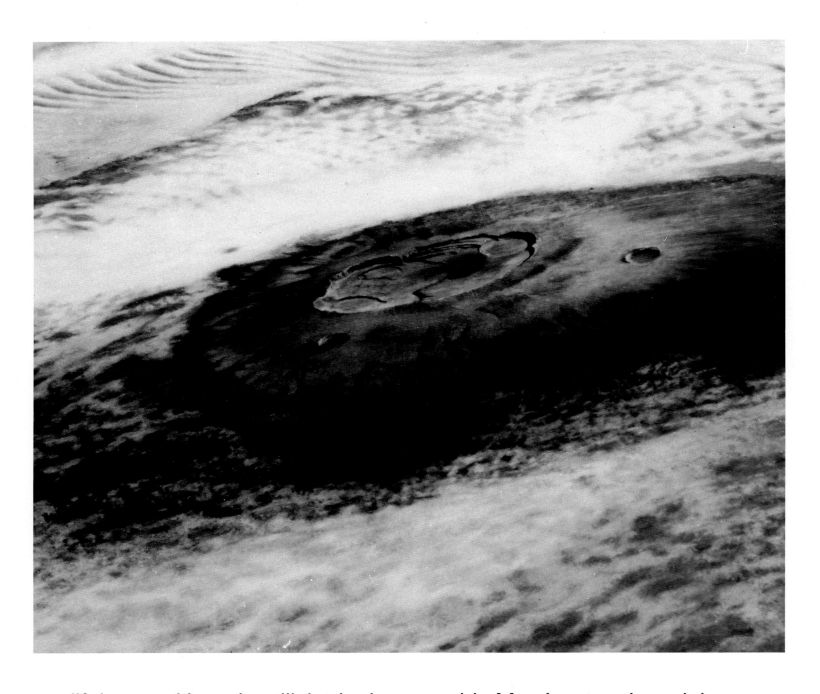

lifeless—although still intriguing—world. Mars' extensive plains contained no green and fertile fields, but there was evidence that a great deal of water once flowed there. The Martian south polar ice cap was found to be half frozen carbon dioxide (dry ice), but the north polar ice cap was found to be almost entirely frozen water.

The details of Martian geography, viewed faintly from Earth, now came into sharp focus. Project Mariner discovered a huge dormant volcano which stood 15.5 miles (25 km) higher than the surrounding plain—three times higher than Mount Everest on Earth—and it was named Mount Olympus after the mythical mountain home of the Greek gods. Also revealed was the true nature of one of Percival Lowell's more prominent 'canals.' What was once thought to be a manmade canal, was actually found to be a vast fracture zone, or canyon, that stretched for 3000

Mars' small diameter (4212 miles) (6739 km) encompasses some enormous surface features. *Above left:* A conception of Olympus Mons—looming 15.5 miles (25 km) above Mars' mean radius, with thrice Mount Everest's height and over 50 times the volume of Mauna Loa, Earth's largest shield volcano. *Above:* A Viking Orbiter image of Argyre Planitia (see caption, page 42). The horizon haze is Mars' carbon dioxide atmosphere.

miles (4800 km) across the Martian equator. Now named Mariner Valley after the project which located it, this canyon, with a width of 125 miles (200 km), is four times deeper than the Earth's Grand Canyon. The dark patches on the Martian surface that seemed to change size seasonally, and which Lowell had interpreted as vegetation, were discovered to be enormous dust storms.

In 1976 the two American Viking spacecraft became the first craft from Earth to actually survive a landing on the Martian surface, and for the first time Earthlings were treated to photographs of the Red Planet from a Martian's eye view. The landing sites chosen for the Viking landers were in the flat lowlands of the northern hemisphere, as these were deemed relatively safe places to land. The Mariner Valley, or the rugged polar regions, would have been more interesting to photograph, but they would have been much riskier sites to land a vehicle that was being operated by remote control from 60 million miles (100 million km) away.

Nevertheless, the landscape was breathtaking. For the first time, men were able to see another planet from ground level. The Viking landers continued to operate long enough to photograph the changing of the seasons, including light snowfalls.

The Viking landers answered a great many questions about Earth's nearest planetary neighbor, but the notion of Martian 'life' remains an enigma. There were three biology experiments aboard the Viking landers that were specifically designed to detect evidence of Martian life, but the answer returned was a resounding 'maybe not.'

Above far left: A Viking Orbiter image of the 3000-mile-long (4800 km) Valles Marineris canyon system, four times as deep as Earth's Grand Canyon.
Above: A Viking Lander 1 view of Mars' Chryse Planitia, in Mars' northwestern quadrant. The rocks' redness evidences hydrous ferric oxide, while the air is reddened by suspended dust. *At top, above,* is a Viking Lander 2 image of the Utopia Planitia, in Mars' northeastern quadrant. This and the image *at above right* evidence Mars' carbon-dioxide-and-water-ice winter snow. See the caption on page 10.

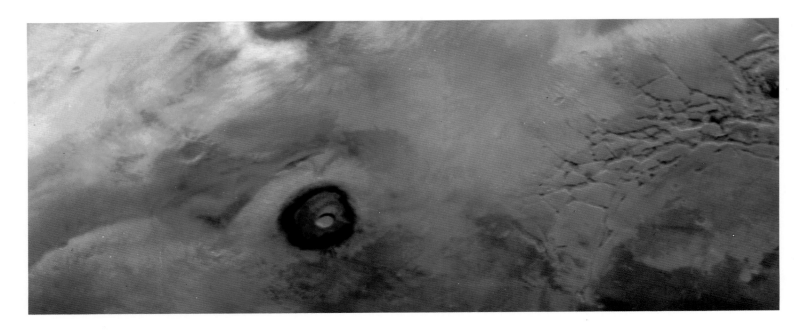

In each experiment, samples of Martian soil were scooped up by the landers' remote surface sample arms and brought aboard the spacecraft. The Pyrolytic Release Experiment was designed to demonstrate whether Martian organisms would be able to assimilate and reduce carbon monoxide or carbon dioxide as plants on Earth do. The easily monitored isotope carbon-14 was used, and the results were described as 'weakly positive.' While the experiment could not be repeated by Viking on Mars, parallel experiments on Earth showed that the same results could *possibly* be explained by chemical, rather than biological, reactions.

In the Labeled Release Experiment, an organic nutrient 'broth' was prepared and 'fed' to some samples of Martian soil, again using carbon-14 as the trace element. It was hypothesized that if microorganisms were present, they would 'breath out' carbon dioxide as they 'ate' the nutrients. Carbon dioxide was, in fact, detected. However, the outgassing of carbon dioxide stopped and could not be restarted. This could have indicated either the existence of some sort of chemical reaction or that a microbe *had* been present but had died while 'eating' the broth.

To distinguish a chemical reaction from a biological reaction, the mixture was heated. This process stopped whatever it was that was producing the carbon dioxide, which *should* have ruled against the notion of a chemical reaction, but which might confirm that it had been caused by a now-deceased organism. In the end, the Labeled Release Experiment was labeled inconclusive because the activities of whatever produced the carbon dioxide had no exact parallel with known reactions of Earth life.

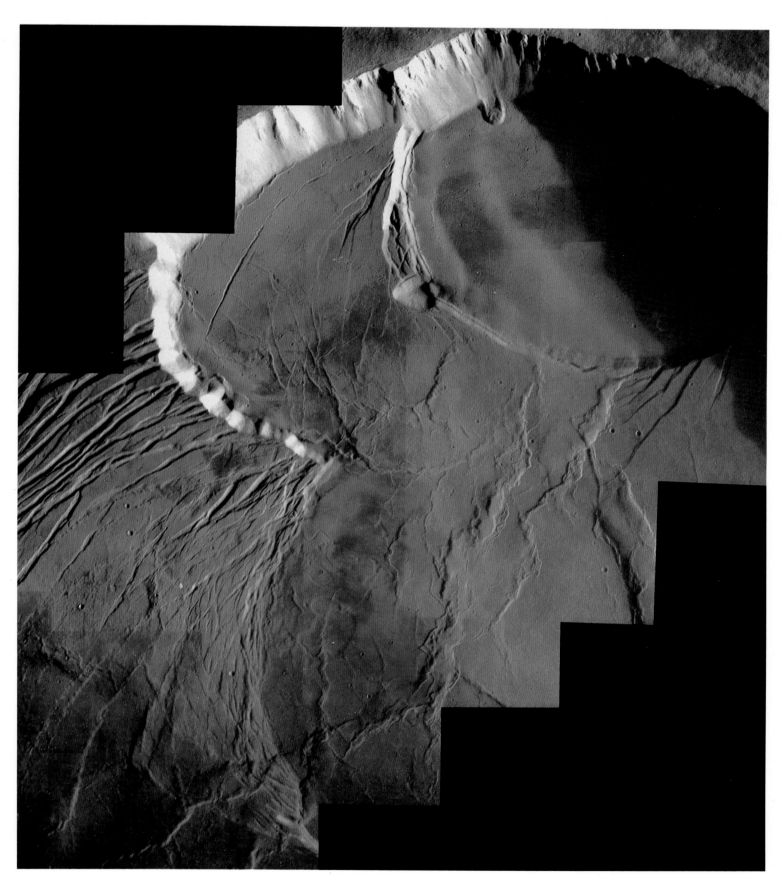

Above far left: A 13 July 1980 Viking Orbiter 1 image of the Tharsis Ridge. Shown are the volcanoes Ascraeus Mons (at top margin) and Pavonis Mons, which rise approximately 10.8 miles (17.3 km) above the ridge, and are just southeast of Olympus Mons, on the Amazonis Planitia, in Mars' northwest quadrant. *Above:* The summit caldera of Olympus Mons, imaged by a Viking Orbiter. This series of overlapping craters is more than 44 miles (70 km) in diameter, and yet is not the largest Martian caldera—Arsia Mons' is 87 miles (140 km) in diameter.

VIKING LANDER 2   CAMERA 2   CE LABEL 220003/000
DIODE BEU/T   STEP SIZE 0.12   CHANNEL/MODE 2/1
VIKING LANDER 2   CAMERA 2   CE LABEL 220016/002
DIODE BEU/T   STEP SIZE 0.12   CHANNEL/MODE 2/1
VIKING LANDER 2   CAMERA 2   CE LABEL 220018/002
DIODE BEU/T   STEP SIZE 0.12   CHANNEL/MODE 2/1
COLOR MOSAIC OF RADCAM OUTPUT SPEC MIN 0. MAX 4.5 *
LABCAT
SAR - LGEOM
MASKVL
                         SEGMENT 1 OF 1
                 IPL PIC ID 76/09/14/125932   WDB/L1473BX
                     JPL IMAGE PROCESSING LABORATORY

The Gas Exchange Experiment was designed to examine Martian soil samples for evidence of gaseous metabolic changes, by again mixing a sample with a nutrient broth. Because the Martian environment is so dry, it was decided to gradually humidify the samples before plunging them into the broth so as not to 'shock' any of the life forms that might be present. A major shock came instead to the Earth-based experimenters as the sample was being humidified—there was a sudden burst of oxygen! When the nutrient broth itself was added, there was some evidence of carbon dioxide, but no more oxygen. Once again the results were described as 'inconclusive' because the results could not be explained by known biological reactions. Subsequent studies have been undertaken in an attempt to determine whether some type of oxidizing agent exists in the Martian soil which could provide a 'chemical reaction' explanation for the strange results of the Gas Exchange Experiment.

The three Viking biology experiments raised some curious questions, but there remains no conclusive answer to the question that has intrigued Earthbound observers. Perhaps the answer lies closer to the Martian poles where there is more water, or perhaps the question might be restated as whether life *might have existed* at one time on Mars. In the long-gone days when rivers ran on the Martian surface, did some civilization—or even just a species of moss—flourish on their banks? Will paleontologists or archaeologists from Earth one day discover fossils or ruins amid the drifting, rust-red Martian sands?

At top, above: An artist's concept of a manned Mars mission. It would take Earth astronauts a year to reach Mars—some 48 million miles (76.8 million km) from Earth at its closest. The image *above,* however, led to light-hearted speculation that we missed being Mars' *first* inhabitants. This illusory, mile-wide (1.6 km) rock formation was imaged by Viking Orbiter I, from 1162 miles (1860 km) out.

   Above far left: A Viking Lander 2 image of Utopia Planitia, with processing data from NASA's Jet Propulsion Laboratories. Parts of the Lander can be seen in the photo, and the black portion at the top is undeveloped.

JUPITER

The largest planet in the Solar System, Jupiter is thought to have once been more than 10 times its present size. In the primeval days during the creation of the Solar System 4.5 billion years ago, Jupiter was a ball of gases that was probably in the process of becoming a *star*, a second—albeit smaller—Sun.

Somehow, however, the nuclear fusion reactions within Jupiter did not become self-sustaining (as had those within the Sun), and Jupiter cooled and collapsed to its current size. Nevertheless, there is still as much energy being radiated from *within* Jupiter as the planet receives from the Sun. Indeed, the great plasma cloud that emanates into space from Jupiter is hotter than the center of the Sun.

Despite never achieving stardom, Jupiter is still an impressive body, the most massive of the planets. With a diameter of 88,650 miles (142,984 km), Jupiter has 1330 times the volume and 318 times the mass of the Earth. Positioned 506 million miles (816 million km) from the Sun, Jupiter appears as the fourth largest object in the sky as viewed from the Earth, next in size after the Sun, Moon and Venus. The Greeks and Romans identified Jupiter as the 'king of planets,' and called it Zeus and Jupiter, respectively, after their principal deities.

Jupiter takes the equivalent of 4333 Earth days to complete a revolution around the Sun. While the Jovian year may be more than 10 times that of the Earth, the Jovian day lasts just 9.84 hours. This rapid rotation, combined with Jupiter's very active atmospheric weather pattern, gives the king of planets a very agitated, violent appearance.

Even though Jupiter has been observed from Earth for centuries, most of what we know about the planet was revealed by the American spacecraft Pioneer 10, Pioneer 11, Voyager 1 and Voyager 2, which made close-up reconnaissance fly-bys between March 1972 and July 1979. The two 1979 Voyager missions traveled nearest to Jupiter and are the

Jupiter, the largest planet in the Solar System, is 88,650 miles (142,984 km) in diameter—one-tenth the Sun's diameter, and more than 10 times the Earth's— with 1318 times the volume of Earth. *At right:* An artist's concept of Voyager 2 with its dish antenna aimed in the (Earthward) direction of the Sun—which is rising over Jupiter, and is between 459 and 505 million miles (734 and 808 million km) distant.

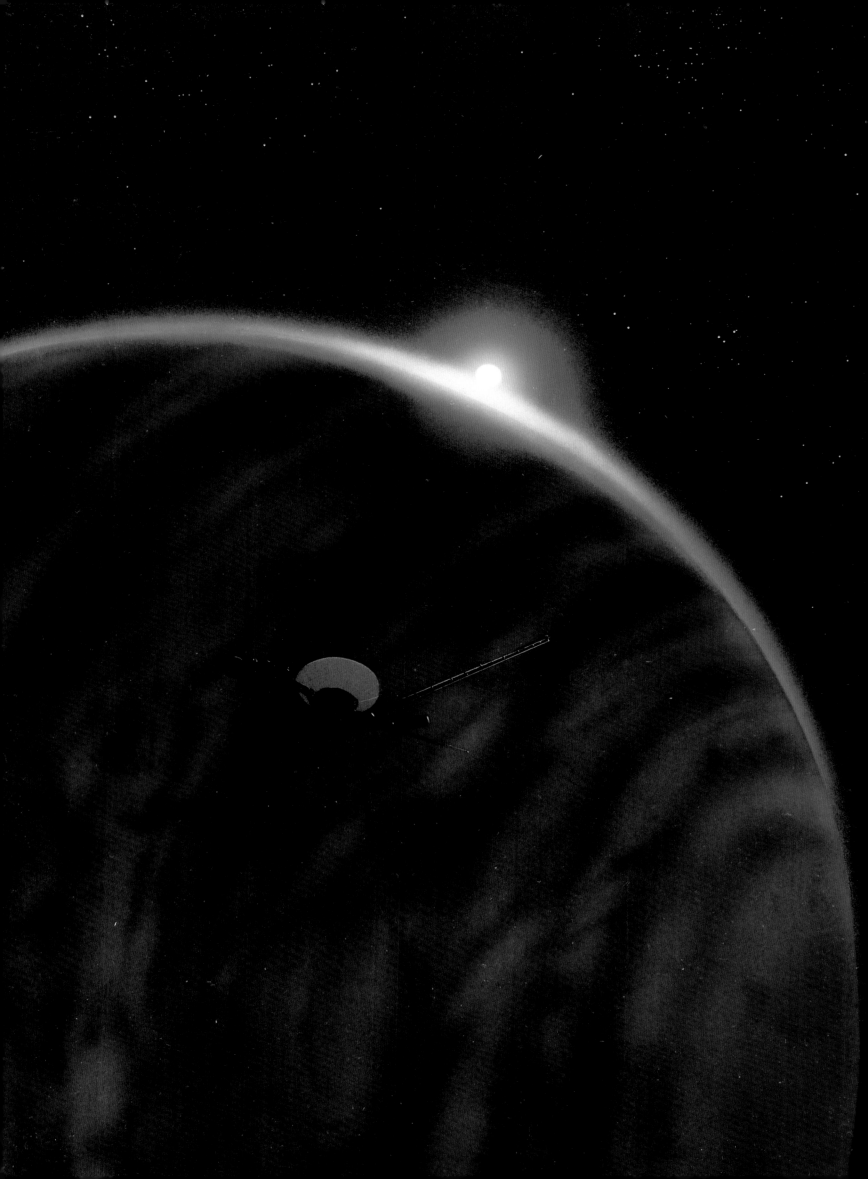

source for most of the photographs in this section. The Galileo space-craft project, launched in 1989, will eventually build upon, and eventually more than double, our present knowledge of turbulent Jupiter.

Unlike the inner, terrestrial planets, Jupiter has neither a solid surface nor molten core, but rather it is gaseous, with a small, solid core. Another, somewhat unorthodox, way of describing Jupiter is to say that it is composed of a sphere of solid silicate rock. Its diameter is about 8000 miles (12,800 km)—the size of the Earth. Upon this sphere there is an ocean of frozen water combined with frozen ammonia that is 4300 miles (6880 km) thick, which is kept frozen by *pressure* rather than *temperature*, because this enormous block of ice averages 37,000 degrees F (15,000 degrees C). Above this incredible glacial sea there is a layer of liquid metallic hydrogen 29,000 miles (46,000 km) deep, topped by a 10,000-mile-deep (16,000 km) ocean of liquid hydrogen. This frothy sea, its waves a boiling, steaming 300 degrees F (150 degrees C), constitutes Jupiter's 'surface,' and over this swirls the planet's turbulent atmosphere.

While most of Jupiter's mass is like that of the Sun— composed of hydrogen and helium—its atmosphere is a noxious mixture which would be highly volatile in an oxygen environment: acetylene, ethane and methane. Also present are ammonia, carbon monoxide, phosphine and even water. The latter, in vapor droplet and ice crystal form, is the primary component of Jupiter's lower cloud layer, while higher clouds consist largely of ammonium hydrosulfide and ammonia.

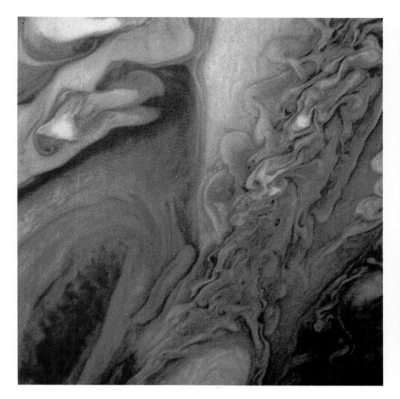

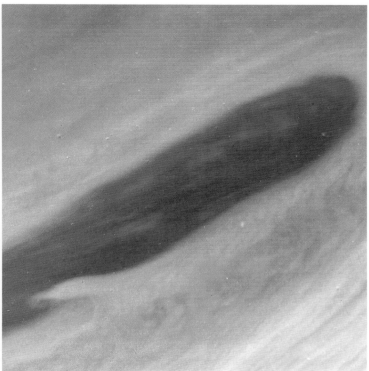

Above far left, both: Voyager 1 and 2 photos show various white spots (or storms) and their migration in respect to the Great Red Spot—which proves independent movement among the atmospheric bands. *Above:* The blue-white band in this view is probably a 'window' into a deeper region of the planet's atmosphere. *At top, above, both:* Closeups of the Jovian atmosphere. *Overleaf:* A Voyager 2 false-color image of Jupiter: above the planet are the satellites Dione (photo right) and Enceladus.

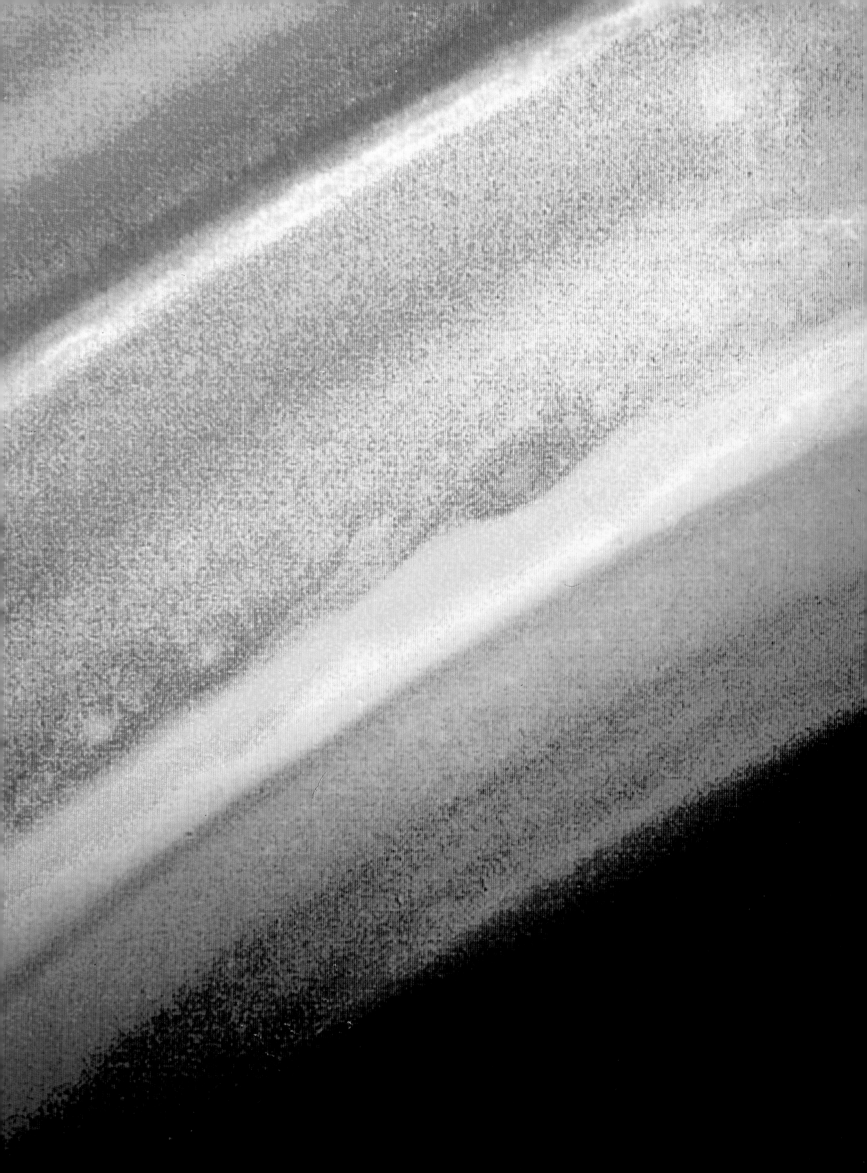

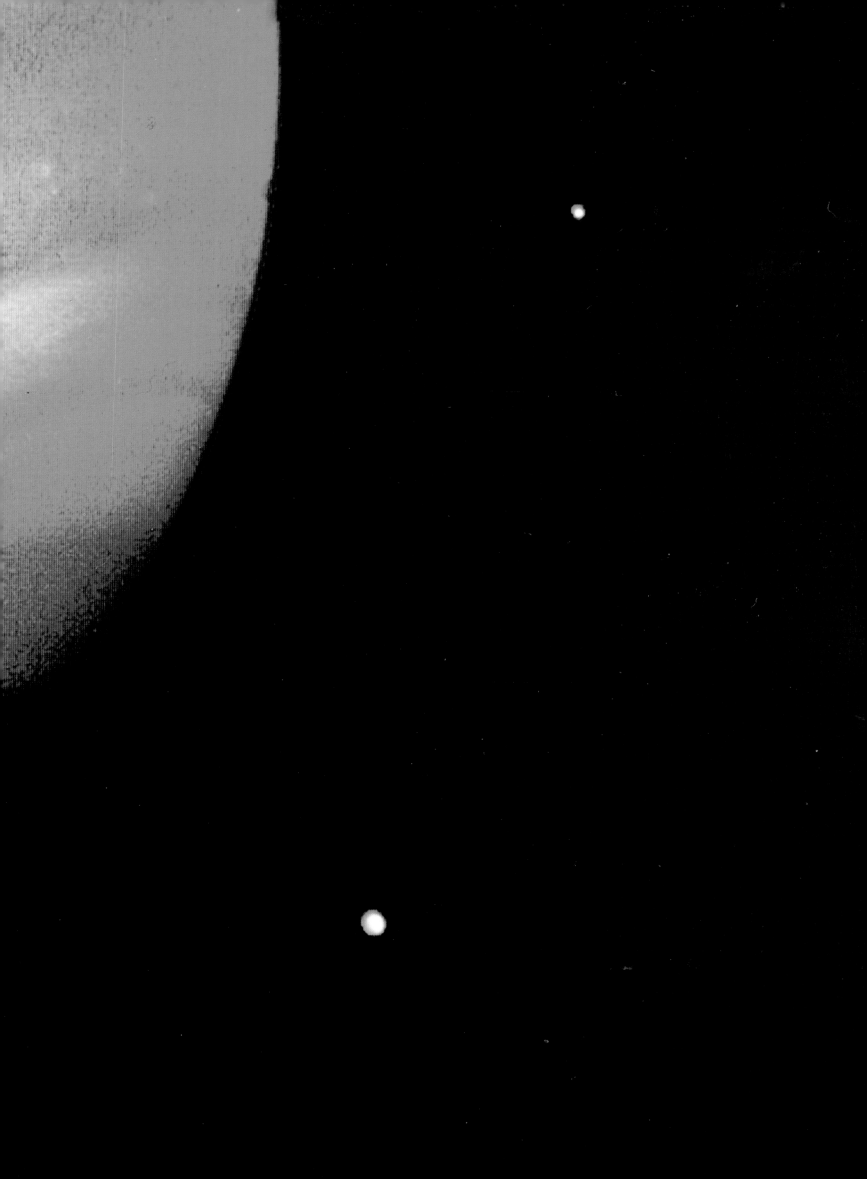

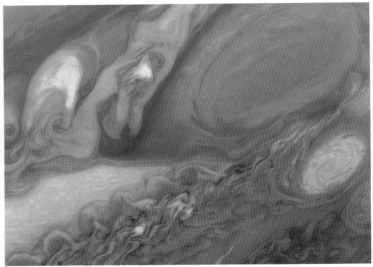

This atmosphere is divided into several distinct parallel bands, which are darker at the poles and in the temperate regions immediately north and south of the planet's equator. Within these undulating bands are numerous elliptical white, brown and red cyclonic storms. The largest of these—and the most outstanding feature on Jupiter—is known as the Great Red Spot (GRS). First recorded in 1664 by the astronomer Robert Hooke, it is a brick red cloud three times the size of the Earth. Described as a 'high pressure system,' the Great Red Spot resembles a storm and exists at a higher and colder altitude than most of Jupiter's cloud cover, although traces of ammonia cirrus are occasionally observed above it. It rotates in a counterclockwise direction, making a complete rotation every six Earth days, and it varies slightly in latitude.

The exact nature of the Great Red Spot is uncertain, but one theory is that it is above an updraft in the Jovian atmosphere in which phosphine—a hydrogen-phosphorus compound—rises to high altitudes, where it is broken down into hydrogen and red phosphorus-4 by solar ultraviolet radiation. The pure phosphorus would give the Great Red Spot its characteristic color.

Another theory pictures the Great Red Spot as the top of a column of stagnant air that exists above an unexplained topographical surface feature far below, within Jupiter. The Great Red Spot is almost certainly the top of some sort of high altitude updraft plume from below the Jovian

Caused by forces far below this cloud layer, atmospheric cloud changes in Jupiter's south equatorial belt (near the Great Red Spot), are evident in the Voyager 1 photo *at top right*, and the Voyager 2 photo *at top left*. *At right*: A Voyager 2 image of Jupiter, from 3.72 million miles (six million km). Time scales on Jupiter are perplexing: the Great Red Spot is a storm that has raged for more than 300 Earth years, while a complete Jovian rotation takes approximately 10 Earth *hours*.

cloud cover, but the divergent flow from it is quite small. For example, one smaller feature was seen to circle the Great Red Spot for an Earth month without altering its distance.

The Great Red Spot may be an awesome feature, but it is a transient one, although any storm that has been raging for over 300 years can certainly be termed an impressive meteorological phenomenon. However, it hasn't been constant in its intensity. Between 1878 and 1882 it was very prominent, but thereafter it dimmed markedly until 1891. Since then it has waned slightly several times—in 1928, 1938 and again in 1977.

Other intriguing meteorological phenomena have also been seen in the Jovian atmosphere, including smaller red spots in the northern hemisphere and some dark brown features that formed at the same latitude as the Great Red Spot. Designated as the South Tropical Disturbance, these features were first observed in 1900, overtook and 'leaped' past the Great Red Spot several times and gradually began to fade in 1935, disappearing five years later.

In 1939 a group of large white spots formed near the Great Red Spot in the southern hemisphere. Like their bigger red counterpart, they rotate counterclockwise. Similar, but smaller, features have been observed in the northern hemisphere, where they are seen to rotate in a clockwise direction. All of these features are oval in shape in the tropical and temperate latitudes, but are more rounded in polar regions. Like the Great Red Spot, the lesser white spots appear to be the tops of some sort of plumes, surrounded by darker filamentary rings.

Jupiter's equatorial band is characterized by regularly spaced features that resemble the type of convective storms that originate in the tropical latitudes on the Earth.

In the northern hemisphere small brown counterclockwise rotating storms race and tumble through the tropical and temperate zones. These features may collide, combine and later break apart.

The most notable interactions between Jovian clouds are in the region of the Great Red Spot. Other spots, or storms, caught on its outer edge might break in two, with one piece remaining in the vortex and the second

At right: A Voyager 2 false-color time exposure of Jupiter's very faint rings. The rings are composed of sand and dust, and are probably less than a mile (1.6 km) thick. As well as rings, Jupiter has 16 moons, ranging from the tiniest, Leda—five miles (eight km) in diameter—to giant Ganymede—3278 miles (5276 km) in diameter—and from the innermost moon, Metis—79,750 miles (127,600 km) from Jupiter—to the outermost, Sinope—14.7 million miles (23.7 million km) from Jupiter. *Overleaf:* A Voyager 2 image of the moon Io, with Jupiter as a backdrop.

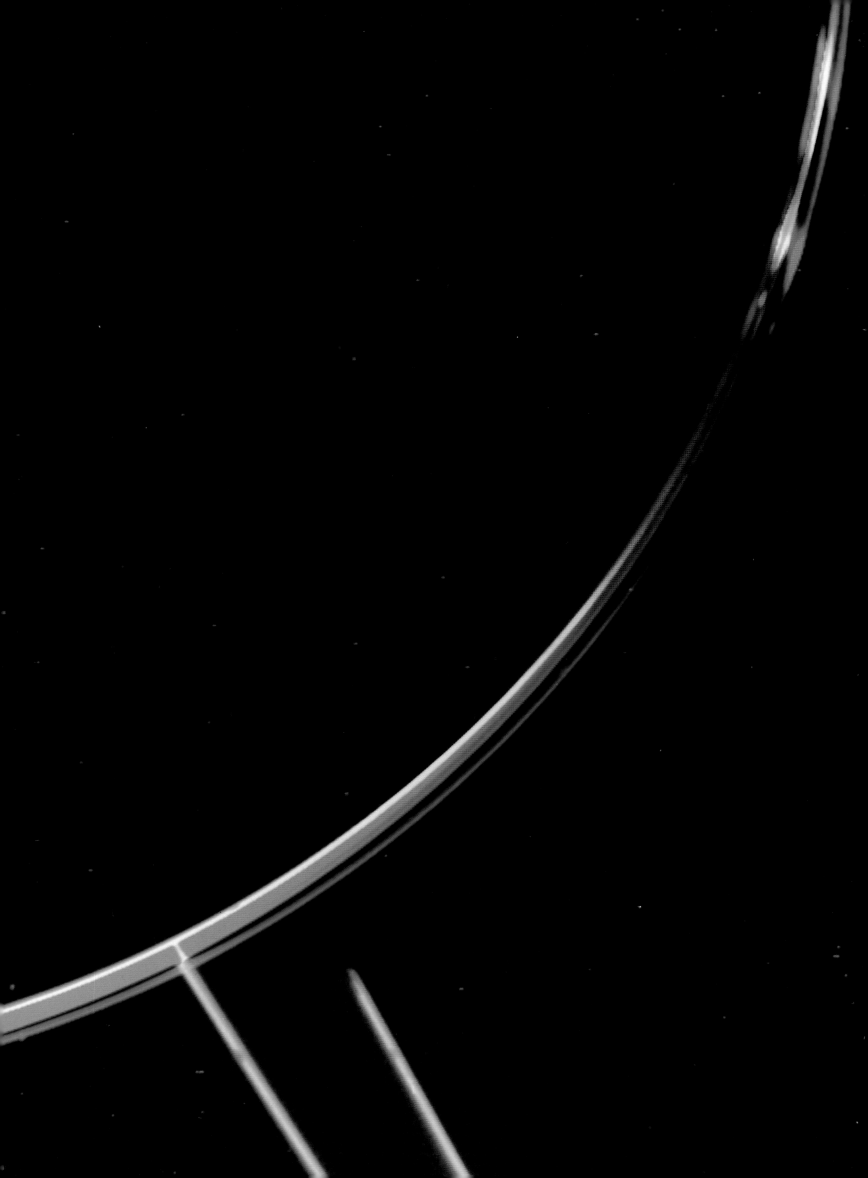

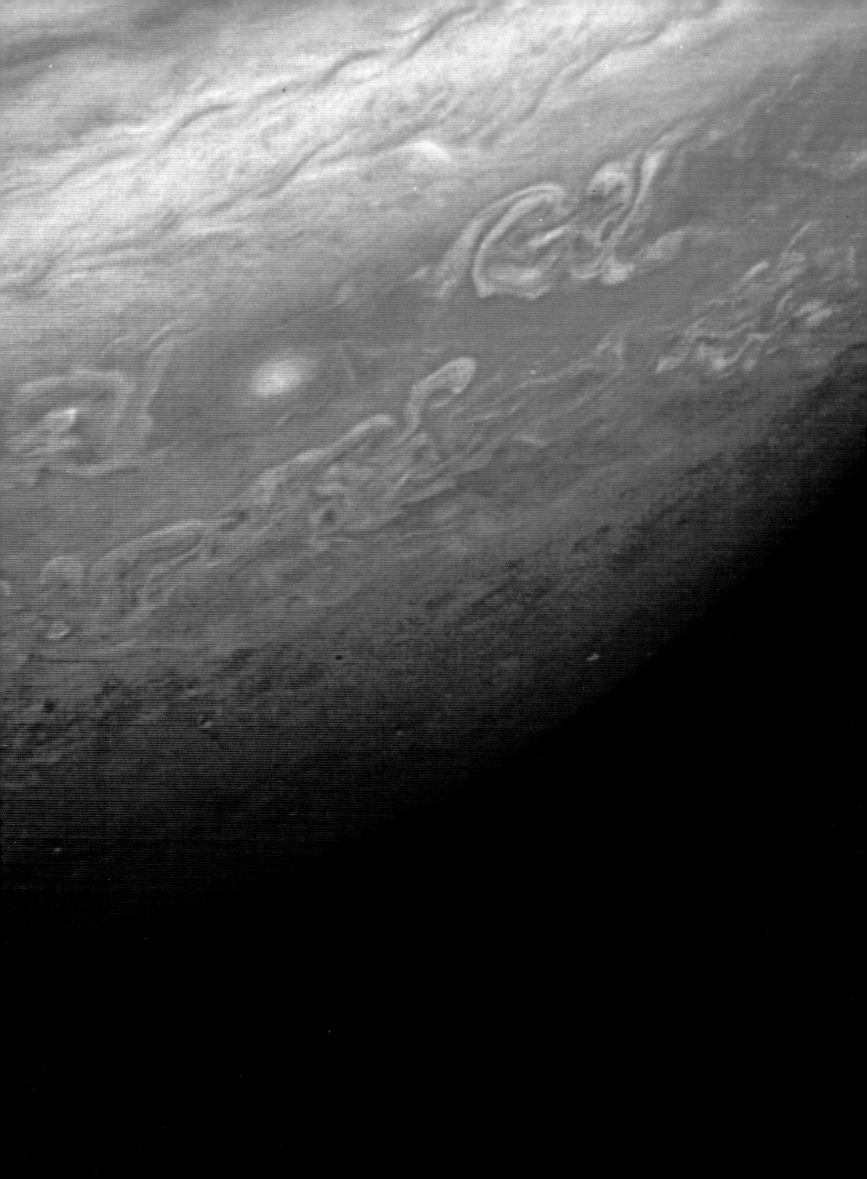

moving away in the same direction as that of the original storm. Occasionally, a ribbon of white clouds might form around the periphery of the Great Red Spot.

In the north polar region there is a very prominent aurora resulting from ultraviolet glows of atomic and molecular hydrogen. Nighttime observations by the Voyager spacecraft also recorded widespread clusters of electrical storms at all latitudes in the Jovian atmosphere.

Unknown before the close-up observations by the Voyager spacecraft in 1979, Jupiter has a distinct ring system. Unlike the very visible Saturnian rings, the Jovian rings are very thin and narrow, and are not visible except when viewed from behind the night side of the planet, when they are backlit by the Sun. The ring system is divided into two parts that begin 29,000 miles (46,000 km) above Jupiter's cloud tops, although some traces of ring material exist below that altitude. The two parts are a faint band 3100 miles (5000 km) across, feathering into a brighter band 500 miles (800 km) across. The rings are composed of dark grains of sand and dust, and are probably not more than one mile (1.6 km) thick.

Jupiter has a large stable of terrestrial moons, but unlike the moon systems of the other giant gaseous planets, all 16 of its known moons are organized into an orderly system of four dissimilar groups, each comprised of four similar sized moons orbiting in distinctly different planes.

The inner group were (except for Amalthea) all discovered during the Voyager project, and they have diameters of less than 200 miles (320 km). They all orbit in a plane whose orbital inclination is less than one-half a degree, and they are located less than 140,000 miles (224,000 km) from Jupiter.

The second group, known as 'The Galileans,' were identified by Galileo Galilei in 1610, and they have diameters greater than 1900 miles (3000 km). Indeed, Ganymede, with a diameter of 3278 miles (5276 km), is the *largest* moon in the *entire* Solar System, and is even larger than the planets of Mercury and Pluto! The Galileans all orbit in a plane whose orbital inclination is less than one-half a degree, and they are all between 250,00 and 700,00 miles (400,000 and 1.1 million km) from Jupiter.

The third group were all discovered in the twentieth century prior to Voyager, and they all have diameters of less than 105 miles (168 km). They all orbit in a plane whose orbital inclination is between 26 and 29 degrees, and they are all between 6.9 and 7.2 million miles (11 and 11.5 million km) from Jupiter.

The Moons of Jupiter

	Discovery	Diameter	Distance from Jupiter
Metis	Project Voyager, 1979	30 mi (49 km)	79,750 mi (127,600 km)
Adrastea	Project Voyager, 1979	21 mi (35 km)	83,030 mi (134,000 km)
Amalthea	Edward Barnard, 1892	103 mi (166 km)	112,655 mi (181,300 km)
Thebe	Project Voyager, 1979	47 mi (75 km)	137,690 mi (222,000 km)
Io	Galileo Galilei, 1610	2257 mi (3632 km)	261,970 mi (421,600 km)
Europa	Galileo Galilei, 1610	1942 mi (3126 km)	416,877 mi (670,900 km)
Ganymede	Galileo Galilei, 1610	3278 mi (5276 km)	664,867 mi (1.1 million km)
Callisto	Galileo Galilei, 1610	2995 mi (4820 km)	1.2 million mi (1.9 million km)
Leda	Charles Kowal, 1974	5 mi (8 km)	6.9 million mi (11.1 million km)
Himalia	CD Perrine, 1904	105 mi (170 km)	7.1 million mi (11.5 million km)
Lysithea	SB Nicholson, 1938	12 mi (19 km)	7.3 million mi (11.7 million km)
Elara	CD Perrine, 1905	50 mi (80 km)	7.3 million mi (11.7 million km)
Ananke	SB Nicholson, 1951	11 mi (17 km)	12.8 million mi (20.7 million km)
Carme	SB Nicholson, 1938	15 mi (24 km)	13.9 million mi (22.4 million km)
Pasiphae	PJ Melotta, 1908	17 mi (27 km)	14.5 million mi (23.3 million km)
Sinope	PJ Melotta, 1914	13 mi (21 km)	14.7 million mi (23.7 million km)

The final group were all discovered in the twentieth century prior to Voyager, and they all have diameters of less than 17 miles. They all orbit in a plane whose orbital inclination is between 147 and 163 degrees, and they are all between 12.8 and 14.7 million miles (20.5 and 23.5 million km) from Jupiter.

Of the Galileans, Europa, Callisto and Ganymede are all described as 'dirty snowballs,' meaning that that they are composed of silicate rock and frozen water. Io, innermost of the Galileans, is totally unlike the others and is one of the most intriguing bodies in the Solar System.

Named for the maiden in Greek mythology who became Zeus' lover and later was turned into a cow by Hera, Io is a unique world. The photographs returned by the Voyager spacecraft in 1979 revealed a very dynamic world, whose surprising characteristics were beyond anything that had previously been imagined. Her surface, with its brilliant reds and yellows that remind one of a giant celestial pizza, is a crust of solid sulfur 12 miles (19 km) thick that floats on a sea of molten sulfur.

This sea is in turn thought to cover silicate rock, which may be partially solid, but is also partially molten. The tidal effect of so massive a body as nearby Jupiter is thought to cause the crust to rise and fall on the molten sulfur by as much as 60 miles (100 km).

It is through this heaving sulfur crust that Io's volcanos have burst. The most volcanically active body in the Solar System, Io is the only place

Above: The sulfurous visage of Io—2257 miles (3632 km) in diameter, and orbiting 261,970 miles (421,600 km) from Jupiter. The doughnut-like feature in the center is a known active volcano. Io is the most volcanically active body in the Solar System, and both lava flows and eruptions of sulfurous material have been documented. *Above far left:* A Voyager 1 view of a volcanic eruption on Io's horizon. Such eruptions loft material as high as 160 miles (256 km) above the moon's surface.

other than Earth and Neptune's moon, Triton, where volcanic eruptions have actually been observed. Io's surface is marred by dozens of jet black volcanos, whose violent eruptions explode up to 130 miles (200 km) into the sky, and surpass (both in magnitude and frequency) anything ever seen on Earth.

In these violent eruptions photographed by the Voyagers, sulfurous material was belched from Io's liquid mantle at speeds of up to 3280 feet per second (1000 meters per second)—many times the recorded velocity of the Earth's volcanos. There were in fact no *small* eruptions recorded among Io's active volcanos on either pass by the Voyagers in March and July 1979, as the smallest plumes all exceeded an altitude of 40 miles (65 km). The extreme altitudes of the plumes and the high velocity of the particles resulted in part from the weaker gravity on Io—whose mass is less than two percent that of the Earth. The sulfur particles also fall back to the surface relatively fast because there is no atmosphere on Io in which they could become suspended, hence no winds to carry them—as was the case on Earth following the eruption of Mount St Helens.

Each of Io's eruptions dumps 10,000 tons (900 metric tons) of sulfur onto the moon's surface. In extrapolation, this would account for 100 billion tons (90 billion metric tons) of sulfur deposits per year. This is enough to bury the entire surface with a layer of sulfur 'ash' one foot thick every 30,750 years. Combined with surface floe, Io could very well be completely resurfaced with a one foot thick layer in as short a time as 3100 years, giving this pizza-colored moon the youngest solid surface in the Solar System aside from Earth, and there are parts of Earth that change less over time than much of Io. In fact, there were many noticeable changes in Io's surface—particularly around Pele—in just the four months between March and July 1979. This 'ever-youthful' surface accounts for the complete absence of meteorite impact craters on Io.

Surrounding the black volcanic caldera are shadowy, fan-shaped features that are the result of liquid sulfur cooling rapidly as it reaches Io's frigid surface. South of Loki, the Voyager imaging team discovered a U-shaped molten lake 125 miles (200 km) across that had partially crusted over. It was detected by its surface crust temperature of about 65 degrees F (20 degrees C)—compared to the surrounding surface temperature of less than −230 degrees F (−110 degrees C). This lake has certainly cooled and solidified by now, while other molten sulfur lakes have no doubt formed elsewhere.

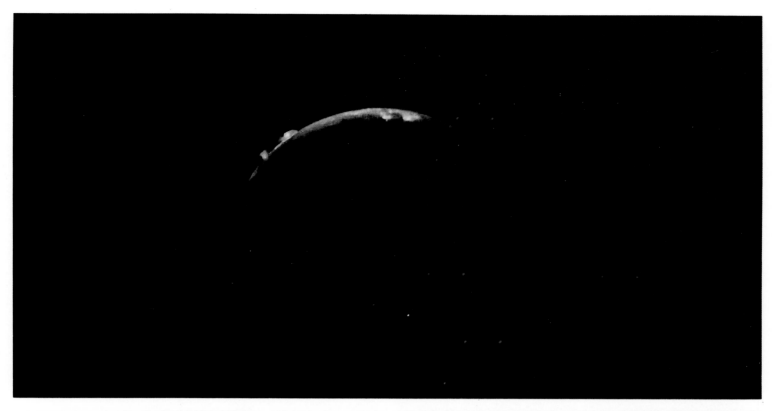

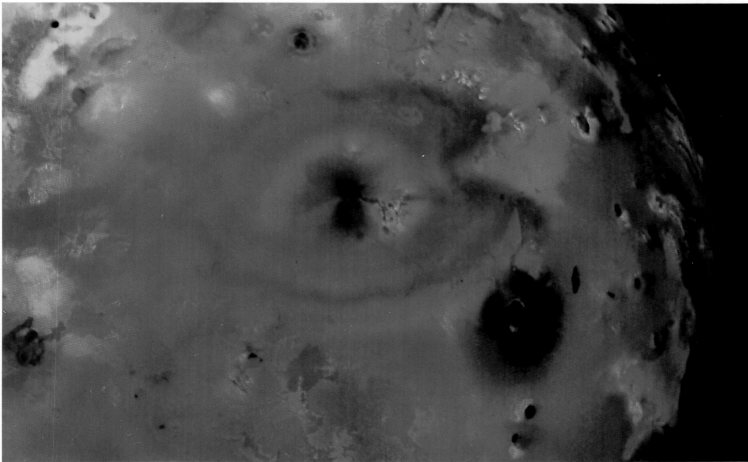

At top, above: A Voyager 2 image of two blue volcanic eruption plumes originating from Io's volcanoes Amirani (upper) and Maui. *Above:* Some prominent
volcanic features on the 'ever youthful' face of Io (see text, this page). From
photo lower right to upper left—Babbar Patera, Mafuike Patera (with the volcano Pele within it), Reiden Patera, Asha Patera, and at photo lower left,
Daedalus Patera. The relatively light area at photo upper left is the Colchis
Regio, a massive sulfur floe.

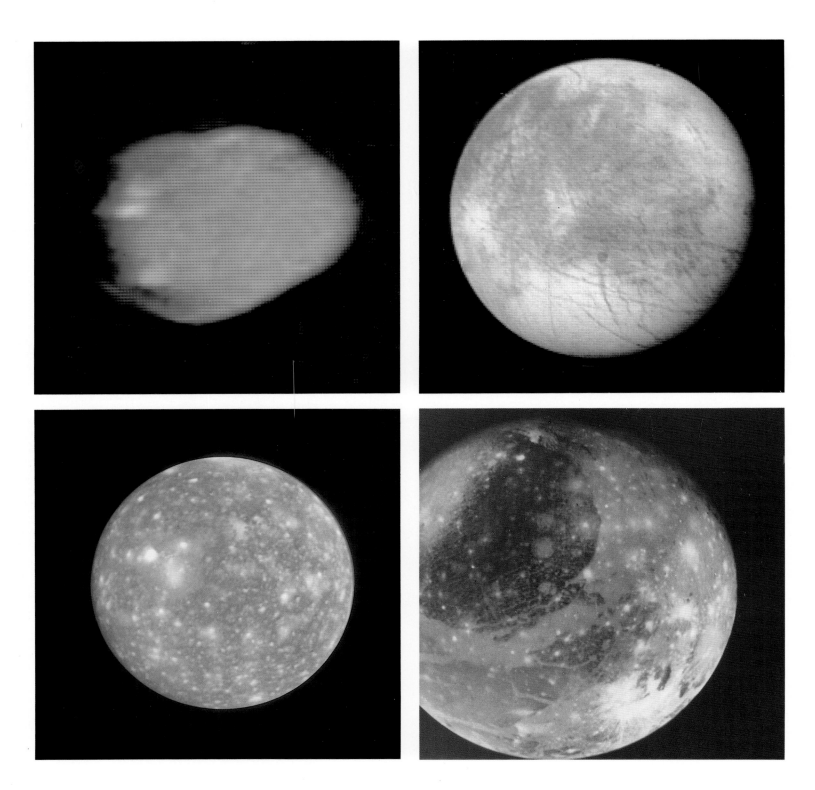

Jupiter's 16 moons are arranged in four groups that are clearly defined as to position and type. The innermost group, Metis, Adrastea, Amalthea *(at top left)* and Thebe, have diameters of less than 200 miles (320 km), have orbital planes of less than half a degree, and orbit less than 140,000 miles (240,000 km) from Jupiter. The largest of these is potato-shaped Amalthea, whose dimensions are 168 x 103 x 93 miles (269 x 165 x 149 km), and whose day equals its orbital period (as is the case with all Jovian moons) at 11.9 Earth hours.

Next out from Jupiter are the Galilean moons Io, Europa *(at top, above)*, Ganymede *(above)* and Callisto *(above left)*—who have diameters greater than 1900 miles (3040 km), orbital planes of less than half a degree and orbits between 250,000 and 700,000 miles (400,000 and 1.2 million km). The largest of these is Ganymede (also the largest moon in the Solar System), 3278 miles (5276 km) in diameter, whose diurnal and orbital periods are 7.16 Earth days. The third group, Leda, Himalia, Lysithea and Elara, have diameters of less than 105 miles (170 km), have orbital planes of between 26 and 29 degrees,

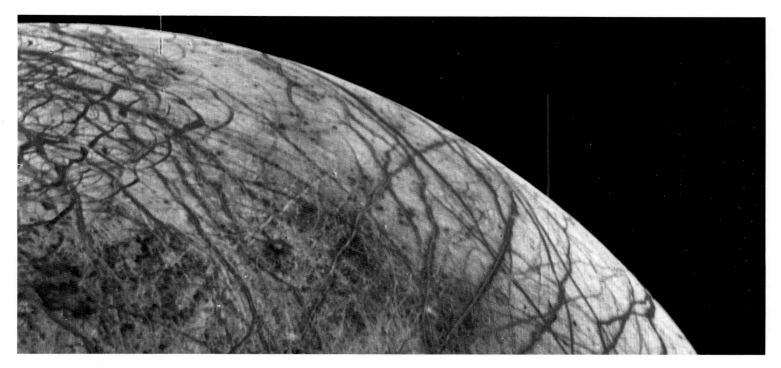

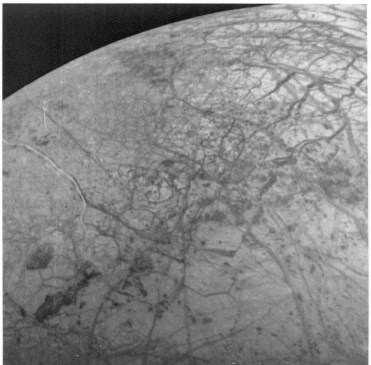

and orbit between 6.9 and 7.2 million miles (11 and 11.5 million km) The largest of these is Himalia, with a diameter of 105 miles (170 km) and diurnal and orbital periods of 250.6 Earth days. The outermost group, Ananke, Carme, Pasiphae and Sinope, have diameters of less than 17 miles (27 km), orbital planes of between 147 and 163 degrees, and orbit between 12.8 and 14.7 million miles (20.7 and 23.7 km) out. The largest of these, Pasiphae, has a diameter of 17 miles (27 km) and diurnal and orbital periods of 735 Earth days.

At top and above: Voyager 2 closeups of Europa, whose surface evidences crack-like lines suggest that darker material has oozed through. The scarcity of impact craters here is a mystery (note those on the other moons). Europa's water-ice crust is 40 miles (64 km) thick, covering an ocean 60 miles (96 km) deep, over a semi-molten silicate core. *Above right:* The dirt-laden, 150-mile-thick (240 km) ice mantle of Callisto. Note the impact craters, and the shatter rings radiating out from 1860-mile-wide (2980 km) Valhalla crater, at photo left.

SATURN

No more awe-inspiring sight exists anywhere than that of the cold visage of Saturn with its distinctive rings. Named for Jupiter's father, the original patriarch of the Roman gods, Saturn was honored by a temple built in 497 BC, which later became the imperial treasury. In 217 BC, the worship of Saturn was conformed to that of its Greek counterpart, Cronus, son of Uranus, and god of boundless time and the cycles. There was a myth that Saturn in Italy, as with Cronus in Greece, had been king during an ancient golden age—and hence was the founder of Italian civilization. Saturn is also associated with the Greek god Phoenon, 'the cruel one,' and the Assyrian god Ninib, patron of agriculture and one of the gods of the Pantheon. From Saturn we have the English word saturnian, or saturnine. Alchemists and early chemists used the name Saturn in reference to its association with the metal lead. Lead poisoning was once called saturnine colic. Astrologers still regard Saturn as possessing an aura of mystery and malevolence.

The most distant planet that is visible when viewed from earth by the unaided eye, Saturn is located an average of 885 million miles (1.4 billion km) from the Sun. It is the second in size of the planets, with a diameter of 74,565 miles (120,000 km). At this distance, it takes Saturn the equivalent of 29.46 Earth years to complete a single revolution around the Sun. In contrast, Saturn's rotational period—its day—is 10.25 hours, less than half that of Earth, but almost identical to that of Jupiter, just as the Martian day is almost equivalent to that of Earth.

Like Jupiter, Saturn has no solid surface. Both planets are thought to have a rocky core about the size of the Earth, which is in turn covered by a layer of water and ammonia that is perpetually sustained in a frozen state by intense *pressure*, despite temperatures that are, on average, hotter than 20,000 degrees F (11,000 degrees C).

Above this icy shell in Saturn's bowels, there rises a liquid metallic hydrogen sea 8500 miles (13,500 km) in depth, which is shrouded by a

Saturn, the second largest planet in the Solar System, is 74,565 miles (120,000 km) in diameter, and has, like Jupiter, an astonishingly short rotational period—10.25 Earth hours. Of course, the most striking feature of this planet are its rings, which were shown by the Voyager spacecraft to be *thousands* of rings. *At right:* A Voyager 1 image of Saturn, its rings and the moon Tethys, whose shadow is at lower right.

layer of 18,500 miles (30,000 km) of liquid hydrogen, with cresting waves that are a chilly − 280 degrees F (−150 degrees C), decidedly colder than Jupiter's surface.

Like Jupiter, Saturn consists primarily of ammonia clouds that swirl through an explosive combination of phosphine, acetylene, ethane, methane, methacetylene and propane. Though these gases are usually considered as flammable, they are, of course, not prone to combustion on Saturn due to the lack of oxygen there.

Saturn's atmosphere, like that of Jupiter, is characterized by separate horizontal bands filled with fiercely swirling storms. These features, while by no means indistinct, are not nearly as pronounced as those on Jupiter. The fact that Saturn's cloud tops are smoother than Jupiter's may be due to weaker gravity (because of smaller mass) and lower temperatures. At these colder temperatures the condensation points of the chemicals in the atmosphere would be reached in regions of higher pressure and hence at lower altitudes within the atmosphere. Thus Saturn's atmosphere is probably characterized by the same visible turbulence as Jupiter's, but at lower altitudes below a layer of ammonia haze.

The most notable feature in Saturn's atmosphere is Anne's Spot, a pale red feature in the southern hemisphere that is thought to be composed of phosphine that is brought high into the upper atmosphere by spiraling convection currents. It is over 3000 miles (4800 km) across, but is a mere speck by comparison with the Great Red Spot of Jupiter.

Saturn's most prominent feature is its vast, complex system of rings. The aggregate diameter of the rings is equal to the distance from the Earth to its Moon, or two and one-half times the diameter of Jupiter!

Galileo Galilei first observed the rings in 1610, but Saturn happened to be oriented so that the great Italian astronomer was viewing them nearly edge-on and thus it wasn't clear what they were. Galileo at first thought he had discovered two identical moons of the scale that he had found near Jupiter. However, these 'moons' did not rotate or change position, and Galileo was mystified. He wrote to the Grand Duke of Tuscany that 'Saturn is not alone but is composed of three, which almost touch one another and never move nor change with respect to one another. They are arranged in a line parallel to the zodiac, and the middle one (*Saturn itself*) is about three times the size of the lateral ones' (*actually the outer edges of the rings*).

By 1612 the plane of the rings was oriented *directly* at the Earth and the

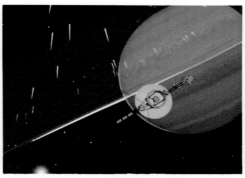

At top, left to right: An artist's conceptual sequence of a Voyager Saturn ring system transit. The bright streaks in the middle view are solid ring particles as observed while speeding past at Voyager velocities—approximately 25,000 mph (40,000 kph). *Above:* A Voyager 1 image of Saturn and moons Mimas and Enceladus, from 47 million miles (75.2 million km). *Overleaf:* Saturn in false color, showing storms (bright spots) welling up in the lower hemisphere's atmospheric bands.

'lateral moons' seemed to disappear entirely. Galileo was completely baffled, but no less so than when they reappeared in 1613. In December 1612 Galileo had written 'I do not know what to say in a case so surprising, so unlooked for and so novel. The shortness of the time, the unexpected nature of the event, the weakness of my understanding, and the fear of being mistaken have greatly confounded me.'

In 1655, more than a decade after Galileo's death, the Dutch astronomer Christiaan Huygens solved the riddle. Using a telescope more powerful than that which was available to his predecessor, Huygens figured out that the mysterious objects were rings around Saturn and the reason for their 'disappearance' in 1612. He also went on to calculate that the rings would be oriented in this way on a 150-month cycle, and that at opposing ends of the same cycle almost the entire ring would be visible from Earth. It has since been determined that the cycle actually alternates between periods of 189 and 165 months. Huygens also discovered Saturn's largest moon, Titan.

In 1671 the Italian-born and naturalized French astronomer Giovanni Domenico (aka Jean Dominique) Cassini began his own observations of the ringed planet. Cassini discovered a second moon of Saturn, Iapetus, in 1671 and four years later he determined that the 'ring' around Saturn was not a single band, but a pair of concentric rings. These two rings would come to be known as the A Ring and B Ring, with the space between them appropriately named the Cassini Division.

In 1837 Johann Franz Encke at the Berlin Observatory tentatively identified a faint division in the A Ring. This division was confirmed in 1888 by James Keeler of the Allegheny Observatory in the United States. Subsequently, this division is known as either the Keeler Gap or (more often) as the Encke Division.

The first spacecraft to venture close to Saturn was the American Pioneer 11 in September 1979. Prior to this time, there were only three known rings of Saturn, each lettered in the order of their discovery from A through C. Pioneer 11 helped Earth-based astronomers identify a fourth ring, which is now known as the F Ring. When the American Voyager Spacecraft first approached Saturn in November 1980, the spectacular

The B Ring is the widest, at 15,800 miles (25,300 km). Saturn's thousands of rings are grouped into seven main sections—the D, C, B, A, F, G and E rings (ascending in order of distance from Saturn)—and two main gaps, or divisions: the Cassini Division (between the A and B rings) and the Encke Division (between the A and F rings). *At right:* The Encke Division, in a computer image from Voyager 2 star occultation data.

photographs they beamed back to Earth revealed that there were not just four, six, or even a dozen rings in Saturn's ring system; rather, there were literally thousands of rings, with each known ring itself composed of hundreds or thousands of rings, with faint rings identified even within the Cassini Division.

In the new nomenclature of Saturn's ring system, the thousands of now known rings and ringlets are divided into seven main rings based on the older nomenclature. Closest to Saturn is a wide, but sparse and extremely faint ring known as the D Ring. At a point roughly 46,000 miles (74,000 km) about Saturn's center and 8700 miles (14,000 km) above Saturn's cloud tops, the D Ring merges into the more distinct C Ring. The C Ring is 10,850 miles (17,400 km) wide, making it the second widest of the easily-visible main rings, though it is less prominent than the slightly narrower A Ring.

At a point 19,600 miles (31,400 km) above Saturn's cloud tops, the C Ring merges into the B Ring without a major gap. The B Ring, which is 15,800 miles (25,300 km) wide, is the brightest of Saturn's main rings and also the widest, except for the virtually invisible E Ring. The 2800 mile-wide (4500 km) Cassini Division separates the B Ring from the 9000 mile-wide (14,400 km) A Ring, the second brightest of Saturn's rings.

The bright, yet tenuous F Ring is located just 2300 miles (3700 km) beyond the outer edge of the A Ring, and the narrow G Ring is located 20,700 miles (33,120 km) from the A Ring. The E Ring is a very, very faint 55,800 mile-wide (89,300 km) mass of particles that begins 91,000 miles (145,000 km) from Saturn's cloud tops and extends past the orbit of the moon Enceladus.

Saturn's rings are composed of silica rock, iron oxide and ice particles, which range from the size of a speck of dust to the size of a small automobile. They range in density from the nearly opaque B Ring to the very sparse E Ring.

Theories about the ring system's origin generally fall into two camps. One theory, originating with Edward Roche in the nineteenth century, holds that the rings were once part of a large moon whose orbit decayed

Saturn's ring system. *Facing page, clockwise from top far right:* The D Ring and its shadow on Saturn's cloudtops; a pre-ring-transit Voyager view of the ring plane; the 'braided' rings of the F Ring; an oblique, false-color view of the ring system (the C Ring is deep blue; the B Ring is burnt sienna to turquoise and the A Ring is bluish white—differences that indicate variations in composition); a false-color view of the rings, from 445,522 miles (712,835 km); and a false-color view of the C Ring, with ringlets in evidence. *Overleaf:* Saturn, with moons Tethys, Dione and Rhea (photo left to right above the planet).

until it came so close to Saturn as to be pulled apart by the planet's tidal or gravitational force. An alternate to this theory suggest that a primordial moon disintegrated as a result of being struck by a large comet or meteorite. The opposing theory is that the rings were never part of a larger body, but rather they are nebular material left over from Saturn's formation 4.6 billion years ago; in other words they were part of the same pool of material out of which Saturn formed, but they remained separate and gradually formed into rings.

Whatever their origin, however, Saturn's rings will continue to distinguish the planet and make it what Galileo described in 1610 as a 'most extraordinary marvel.'

Saturn's moon system is probably the largest and most complex in the Solar System. Whereas Jupiter has 16 moons neatly ordered into four groups of four generally homogenous moons, Saturn has a far flung, and even chaotic, family. Prior to the 1977 launch of the two Voyager spacecraft, Saturn was thought to have nine moons. After the Voyager encounter with the planet in 1980 and 1981, the number of known moons had doubled, and today, we're still not sure that we've seen them all. The main area of confusion lies, predictably, in the region of Saturn's rings, where tiny moons are hidden within millions of square miles, or kilometers, of rock and ice particles.

The Voyager missions confirmed five moons within the diameter of the G Ring, which ranged in size from 12 miles (20 km) to 137 miles (220 km). Of these, Janus and Epimetheus were seen as being in the *same orbit*, and on a possible collision course. Because no spacecraft has visited Saturn since 1981, no one is certain whether this collision took place.

Within a distance of 350,000 miles (560,000 km) of Saturn's cloud tops there are also five moons that were known prior to 1977, and at least three of these—Mimas, Tethys and Dione—each have one or two 'co-orbital' moons. It is even theorized that there may be additional co-orbital moons in this region, and indeed, past collisions may have taken place.

Within the orbit of the moon Rhea, 328,000 miles (525,000 km) from Saturn, there are at least 15 moons. Beyond this environment there are but four known Saturnian moons. These are Titan, at 760,000 miles

The Voyager spacecraft revealed that Saturn had at least 20 moons (including co-orbitals), the innermost of which are involved with the planet's extensive ring system. *Overleaf:* The moon Enceladus, composed mainly of water-ice and rock—its surface grooves are probably the result of crustal fracturing, and its smooth plains may be the product of water welling up from within the moon and freezing on its surface.

The Moons of Saturn

(Excludes Co-orbitals. See text, facing page.)

	Discovery	Diameter	Distance from Saturn
Atlas	Project Voyager, 1980	19 mi (30 km)	85,544 mi (137,670 km)
Prometheus	Project Voyager, 1980	137 mi (220 km)	86,589 mi (139,353 km)
Pandora	Project Voyager, 1980	56 mi (90 km)	88,048 mi (141,700 km)
Epimetheus	Project Voyager, 1980	40 mi (65 km)	94,089 mi (151,422 km)
Janus	Project Voyager, 1980	60 mi (95 km)	94,120 mi (151,472 km)
Mimas	William Herschel, 1789	242 mi (390 km)	115,326 mi (185,600 km)
Enceladus	William Herschel, 1789	311 mi (500 km)	147,948 mi (238,100 km)
Tethys	Giovanni Cassini, 1684	652 mi (1050 km)	182,714 mi (292,342 km)
Telesto	Project Voyager, 1980	9 mi (15 km)	183,118 mi (294,700 km)
Calypso	Project Voyager, 1980	9 mi (15 km)	217,480 mi (350,000 km)
Dione	Giovanni Cassini, 1684	696 mi (1120 km)	234,567 mi (377,500 km)
Helene	Project Voyager, 1980	20 mi (32 km)	234,915 mi (378,060 km)
Rhea	Giovanni Cassini, 1672	951 mi (1530 km)	327,586 mi (527,200 km)
Titan	Christiaan Huygens, 1655	3200 mi (5150 km)	759,067 mi (1.2 million km)
Hyperion	GP Bond and William Lassell, 1848	155 mi (250 km)	921,493 mi (1.5 million km)
Iapetus	Giovanni Cassini, 1671	905 mi (1460 km)	2.2 million mi (3.6 million km)
Phoebe	William Pickering, 1898	137 mi (220 km)	8 million mi (13 million km)

(1.2 million km); Hyperion, at 922,000 miles (1.5 million km); Iapetus, at 2.2 million miles (3.6 million km); and Phoebe, an astonishing eight million miles (13 million km) from Saturn.

Saturn's moons are generally small, with diameters of 950 miles (1530 km) or less, with one exception. Titan is larger than either Mercury or Pluto. Once thought to be the Solar System's largest moon, it is now known to rank a close second to Jupiter's Ganymede.

Prior to the confirmation in 1989 of an atmosphere on Neptune's Triton, Titan was the only known moon with a fully developed atmosphere that consists of more than simply trace gases. It has in fact a denser atmosphere and cloud cover than either Earth or Mars. This cloud cover, nearly as opaque as that which shrouds Venus, has prevented the sort of surface mapping that has been possible with the other major moons of the outer Solar System, but its presence has only served to make Titan all that much more intriguing. (Early astronomers mistook this dense

Above left: Among the thousands of impact craters on the Saturn moon Dione, we see the large impact basins (photo left to right) Dido, Remus and Romulus (overlapping) and Aeneas. The largest of these, Aeneas, is less than 62 miles (99 km) in diameter. The moon's largest crater, Amata, is nearly 150 miles (240 km) across. Dione, like Mimas, Enceladus and Tethys, is a mostly water-ice 'dirty snowball.' It is 696 miles (1120 km) in diameter, and orbits 234,567 miles (377,500 km) out from Saturn. Also, like the smaller Mimas and Tethys (see the chart on page 85), Dione has at least one co-orbital moon, and may have a second. A footnote on the moon Mimas is that it bears a surface feature that is one of the *proportionately* largest in the Solar System—Crater Herschel, at 81 miles (130 km) in diameter, is approximately one-third the diameter of Mimas itself.

Above: An artist's concept of the view of Saturn from the surface of the moon Rhea. Rhea is the second largest moon of Saturn, with a diameter of 951 miles (1530 km), and is composed of equal parts of water-ice and silicate rock. The moon's leading hemisphere is solid ice. It is theorized that this moon's exceptional brightness may be the product of upwellings of ice from the interior—a 'repaving' that renews some parts of the moon's surface from time to time.

Unlike Saturn's largest moon, Titan, Rhea has no atmosphere—hence the crystal clarity of the view proposed in this artwork. Were this a conceptual view from Titan, one would not likely see Saturn at all (see text, this page through page 92). Rhea bears similar wispy surface detail to that observed on Dione: this is the result of ice flooding and encrustation. Titan, on the other hand, is like Venus in that it reveals no surface detail to the observer. Unlike Venus, Titan's atmosphere is mostly nitrogen, with hydrocarbons.

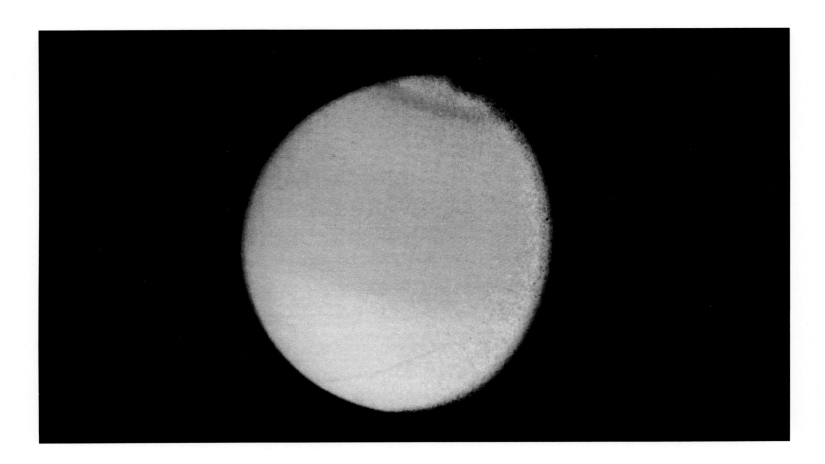

atmosphere for Titan's actual surface, and it was through this mistake that Titan was once considered to be the Solar System's largest moon.)

Titan's atmosphere is extremely rich in nitrogen, the same element that makes up the greatest part of the Earth's atmosphere. Other major components of Titan's atmosphere are hydrocarbon gases, such as acetylene, ethane and propane, with methane being the most common of the hydrocarbons. While these gases are also found in Saturn's atmosphere, Titan's atmosphere contains four times the ppm concentration of ethane and 150 times the concentration of acetylene. Titan's atmosphere probably includes broken methane clouds at an altitude of about 25 miles (40 km), with the dense, smoggy hydrocarbon haze stretching up to an altitude of nearly 200 miles (320 km), where ultraviolet radiation from the Sun converts methane to acetylene or ethane.

The view from Titan's surface is one of an exciting, but inhospitable, world. Covered by the opaque haze, the sky would appear like a smoggy sunset on Earth or like a view from the surface of Venus. The atmospheric pressure on Titan's surface, while 1.6 times that of Earth is,

Titan was the first moon found to have an atmosphere. *At top, above:* A Voyager image of Titan's clouded atmosphere—probably a global aerosol rather than the convective clouds seen on Jupiter and Saturn. *At right:* A Voyager 1 false-color image of Titan, showing (left to right) the upper level of the aerosol; a division at 310 miles (500 km) altitude; a division at 233 miles (375 km); and a division at 124 miles (200 km).

 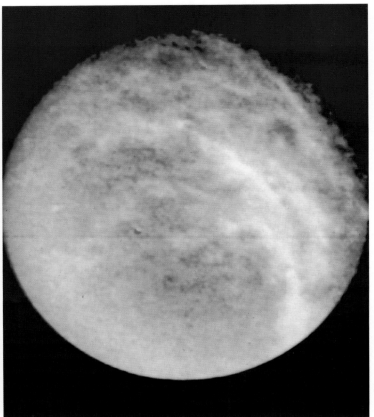

however, a good deal less than that of Venus. Titan's surface temperature of nearly -300 degrees F (-185 degrees C) would permit methane to exist not only as a gas, but also as a liquid or a solid, in much the same way that water does on Earth. A picture is thus painted of a cold, orange-tinted land where methane rain or snow falls from the methane clouds and where methane rivers may flow into methane oceans dotted with methane icebergs.

There is evidence of a 30 Earth-year seasonal cycle which *may* have permitted the development of methane ice caps that expand and recede like the water ice caps on Earth (and the water/carbon dioxide ice caps on Mars). Water ice is also present on Titan, beneath the methane surface features, and possibly extends up into the atmosphere in the form of ice mountains. Titan's mantle is, in turn, largely composed of water ice that gives way to a rocky core perhaps 600 miles (960 km) beneath the surface. The absence of a magnetic field indicates that Titan has no significant amount of ferrous metallic minerals in its core.

The presence of nitrogen, a hydrocarbon atmosphere and water indicate that Titan's surface is very much like that of the Earth four billion years ago before life evolved on the latter body. It has been suggested that this similarity to the prebiotic 'soup' that covered the Earth in those bygone days could presage a similar chain of events on Titan.

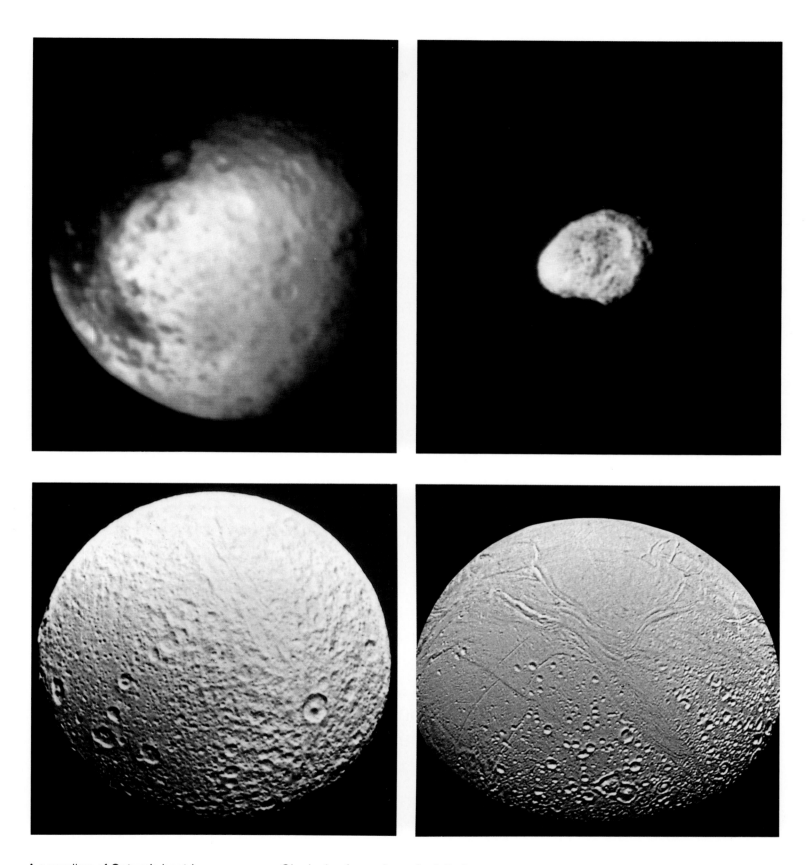

A sampling of Saturn's best-known moons. *Clockwise from above far left:* A Voyager 1 image of Mimas and Crater Herschel (see also caption, page 89); Rhea, evidencing frost and indeterminate surface detail; a Voyager 2 image of Iapetus, third largest of Saturn's moons (905 miles [1460 km] in diameter), with an area of extruded dark subsurface material, Cassini Regio, at photo left; a Voyager 2 image of Hyperion, oddly assymetrical (perhaps collision damaged) for a body of its size—249 x 155 x 149 miles (348 x 248 x 238 km); a Voyager 2 view of the wrinkled face of Enceladus—in evidence are the complex rift systems of Samarkand Sulci and Harran Sulci, backed by the smooth areas known as Sarandib and Diyar *planitiae*; and the heavily cratered and canyoned face of Tethys.

URANUS

For many centuries, mankind lived by a celestial mythology constructed around the comings and goings of the five planets that were then known. Even today many orthodox astrologers refuse to attribute any magical significance to the planets beyond Saturn. Against this backdrop, what a stunning discovery it must have been when the great astronomer William Herschel discovered a new planet in 1781.

Located in an orbit more than twice as far from the Sun as Saturn, Uranus takes 84 Earth years to complete a revolution around the Sun. It is named for the earliest supreme god of Greek mythology. The personification of the sky, mythical Uranus was both son and consort to the goddess Gaea and father of all the Cyclopes and Titans.

Uranus is a gaseous planet like Jupiter, Saturn and Neptune, with a distinct blue-green appearance, probably due to a concentration of methane in its upper atmosphere. In terms of size, it is smaller than Jupiter and Saturn, while being very close to the size of Neptune. Its solid core is composed of metals and silicate rock with a diameter of roughly 32,116 miles (51,800 km). Its core is, in turn, covered by an icy mantle of methane ammonia and water ice 6000 miles (9600 km) deep.

As with the other gaseous planets, the predominant elements in the Uranian atmosphere are hydrogen and helium, although the Voyager 2 observations in 1986 indicated that the atmosphere was only 15 percent helium, versus 40 percent, as originally postulated. Other atmospheric constituents include methane, acetylene and other hydrocarbons.

The clouds that form in this atmosphere are moved by prevailing winds that blow in the same direction as the planet rotates, just as they do on Jupiter, Saturn and Earth.

Uranus has an axial inclination of 98 degrees. Discovered in 1829, this is a phenomenon that is unique in the Solar System. With such an axial inclination, Uranus is seen as rotating 'on its side' at a near right angle to

At right: A Voyager 2 view of Uranus on 25 January 1986, with the Uranian crescent backlit by the Sun—which is 1.7 to 1.9 billion miles (2.7 to three billion km) distant. Unlike other known planets in the Solar System, Uranus rotates at a 90-degree angle—presenting its poles, rather than its equatorial regions, to the Sun. This gas giant's diameter is 32,116 miles (51,800 km), and its day is 17.3 Earth hours long.

the inclination of the Earth or Sun. The poles of Uranus, rather than its equatorial regions, are pointing alternately at the Sun. This rotation is equal to 17.3 hours, which is slower than either Jupiter or Saturn but still much faster than that of the Earth.

Prior to the flyby of Voyager 2 in January 1986, Uranus was thought not to have a magnetic field, but this assumption proved false. The magnetic field of Uranus is tilted at a 60 degree angle to the planet's rotational axis (compared to 12 degrees on Earth). The magnetic field has roughly the same intensity as the Earth's, but whereas the Earth's magnetic field is generated by a molten metallic core, the one surrounding Uranus seems to be generated by the electrically conductive, super pressurized ocean of ammonia and water that exists beneath the atmosphere.

Uranus, like Jupiter and Saturn, has a system of rings, of which the first nine were discovered by Earth-based observers in 1977. In 1986 Voyager 2 observed these in detail and identified two more. This ring system

The Uranian rings are dark, and hence cannot be readily seen. Stellar occultation is used to discern their quantity and extent. *At top, above:* A false-color Voyager portrait of Uranus' rings (photo top to bottom): the Episilon, Delta, Gamma, Eta, Beta, Alpha, Four, Five and Six rings are seen here: 1986U2R is not seen. See the text, pages 96—98. *At right:* False-color sections of the Epsilon Ring, showing its size fluctuations.

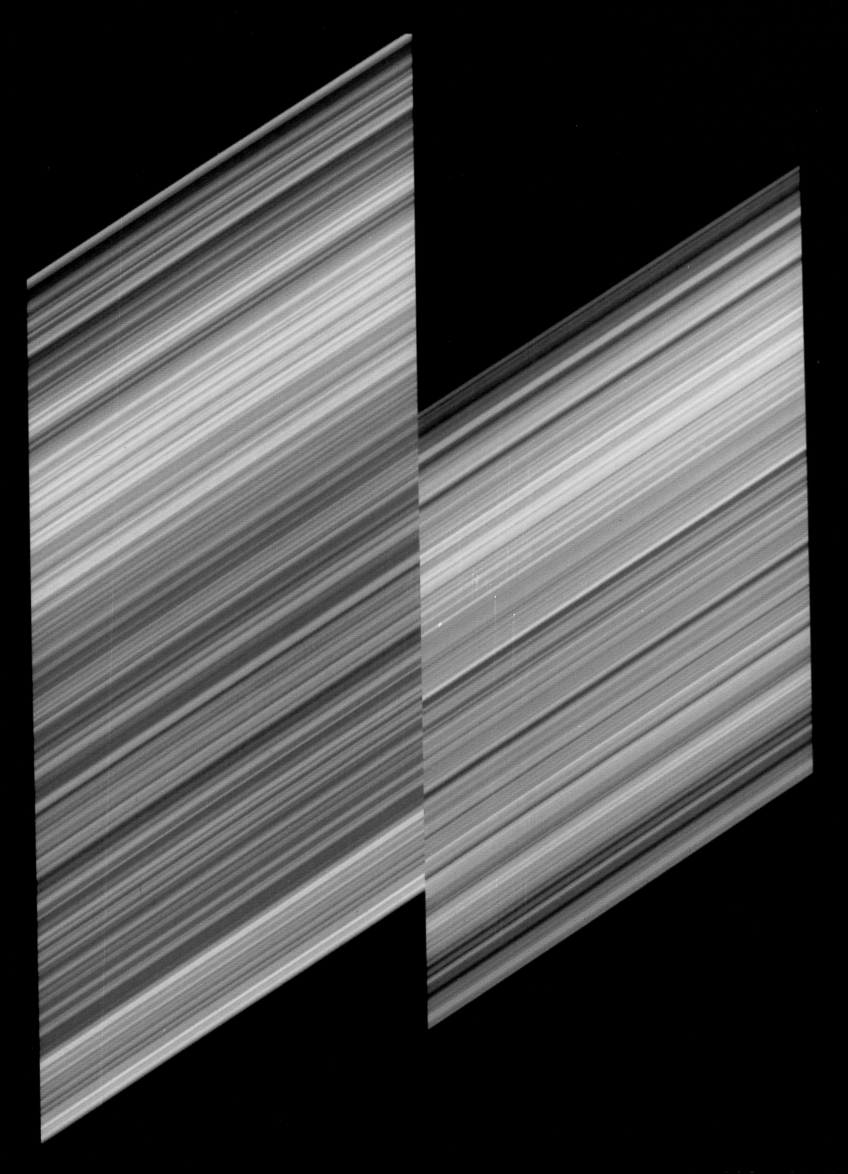

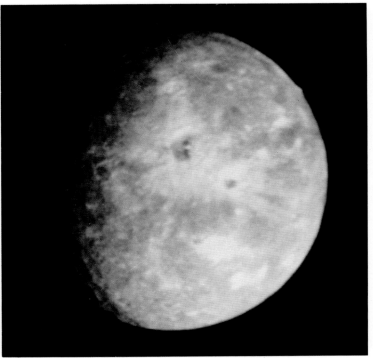

is much more complex than Jupiter's, but less so than Saturn's spectacular system. The system around Uranus seems to be relatively young and probably did not form at the same time as the planet. The particles that make up the rings may be the remnants of a moon that was broken by a violent impact or torn apart by gravitational effects of Uranus.

The widest ring known before Voyager 2 was the outermost ring, Epsilon—an irregular ring measuring 14 to 60 miles (22 to 96 km) across. In 1986 Voyager's cameras helped identify a new innermost ring, designated 1986U2R, which is 1550 miles (2500 km) wide. The narrowest complete rings are less than a mile wide.

Prior to the observations by Voyager 2, Uranus was known to have just five moons. Photographs returned by the spacecraft increased the number of known moons to 15, with all 10 of the newly-discovered moons located *within* the orbital paths of the original five. One of the new moons, Puck (1985U1), was discovered by Voyager's cameras in late 1985, and the rest were discovered in the photos taken during the January 1986 Voyager flyby of the Uranian system. With the exception of Puck and Cordelia—the largest and smallest of the 'Voyager' moons—all of the

Titania and Oberon are the largest Uranian moons, with diameters of 998 and 961 miles (1610 and 1550 km), respectively, and are also are the farthest moons from the planet, at 271,104 and 326,507 miles (436,300 and 583,400 km), respectively. Note the Voyager 2 views of Titania *(at top, left)* and Oberon *(at top, right)*. As with Jupiter, Uranus' moons are grouped with uncanny orderliness—here, with an almost perfect correlation between size and orbital distance from Uranus. See the chart on page 99.

The Moons of Uranus

	Discovery	Diameter	Distance from Uranus
Cordelia	Project Voyager, 1986	25 mi (40 km)	30,882 mi (49,700 km)
Ophelia	Project Voyager, 1986	31 mi (50 km)	33,429 mi (53,800 km)
Bianca	Project Voyager, 1986	31 mi (50 km)	36,785 mi (59,200 km)
Juliet	Project Voyager, 1986	37 mi (60 km)	38,400 mi (61,800 km)
Desdemona	Project Voyager, 1986	37 mi (60 km)	38,959 mi (62,700 km)
Rosalind	Project Voyager, 1986	50 mi (80 km)	40,140 mi (64,600 km)
Portia	Project Voyager, 1986	50 mi (80 km)	41,072 mi (66,100 km)
Cressida	Project Voyager, 1986	37 mi (60 km)	43,433 mi (69,900 km)
Belinda	Project Voyager, 1986	37 mi (60 km)	46,789 mi (75,300 km)
Puck	Project Voyager, 1985	106 mi (170 km)	53,437 mi (86,000 km)
Miranda	Gerard Kuiper, 1948	217 mi (150 km)	80,716 mi (128,282 km)
Ariel	William Lassell, 1851	721 mi (1160 km)	118,358 mi (190,900 km)
Umbriel	William Lassell, 1851	739 mi (1190 km)	165,284 mi (266,000 km)
Titania	William Herschel, 1787	998 mi (1610 km)	271,104 mi (436,300 km)
Oberon	William Herschel, 1787	961 mi (1550 km)	326,507 mi (583,400 km)

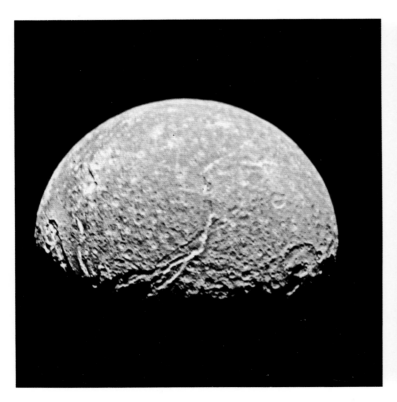 

newly discovered members of the group are very uniform in size, with diameters ranging between 31 and 37 miles (50 and 59 km).

The innermost of the moons are Cordelia (1986 U7), located between the Delta Ring and the Epsilon Ring, and Ophelia (1986 U8), on the opposite side of the Epsilon Ring. Thus straddling the Epsilon ring, these two small bodies probably act like the shepherd moons of Saturn, controlling and defining the position and shape of the ring.

The five largest of the Uranian moons range in size from the 217 miles (150 km) diameter of little Miranda, to the 998 miles (1610 km) diameter of Titania. Like most of Jupiter's Galilean moons and many of Saturn's mid-sized moons, the 'big five' Uranian moons are 'dirty snowballs,' ie, spheroids composed mainly of frozen water and silicate rock, with lesser amounts of methane ice.

Ariel, Umbriel and Miranda are mostly water-ice, with some methane-ice and a core of silicate rock. *At top, left:* Ariel, the brightest Uranian moon, showing, at photo left, Kachina Chasma, and at photo right, the complex of rifts that include Korrigan, Pixie and Brownie chasmas. *At top, above:* Umbriel, evidencing rift canyons and heavy cratering.

Above right: A Voyager 2 closeup of Miranda, showing (photo left to right) the rugged terrain of Silicia Regio, the rocky valley called Verona Rupe, the tip of the pressure fault system known as Inverness Corona, and Crater Alonso. *At right:* Elsinore Corona and Inverness Corona.

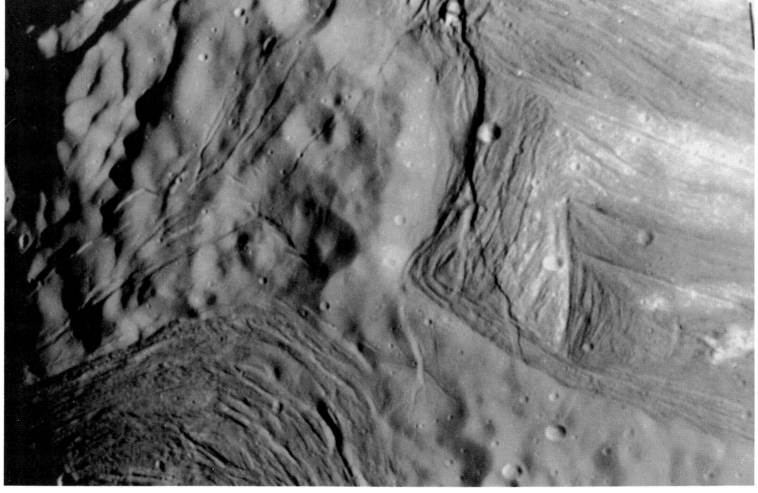

NEPTUNE

After the discovery of Uranus, it took but 65 years before the existence of an eighth planet was confirmed. Galileo had sighted this object as early as 1613, but it was Johann Galle and Heinrich Ludwig D'Arrest who finally identified it as a planet in 1846. They named it Neptune, after the Roman sea god, because of its pale, sea-green color.

Located 2.8 billion miles (4.5 billion km) from the Sun, Neptune is three times more distant from the Sun than Saturn and more than half again farther out as Uranus. At this distance, it takes 165 years for Neptune to complete one revolution around the Sun. Like Jupiter, Saturn and Uranus, Neptune is a giant gaseous orb. With a diameter of 30,642 miles (49,424 km) Neptune is a near twin of Uranus and is as close in size to this neighbor as Venus is to the Earth. Like Uranus, Neptune has a longer rotational period—16.3 hours—than Jupiter or Saturn, yet this rotational period is less than that of the terrestrial planets. Neptune also corresponds to Uranus in terms of its physical composition, which consists primarily of hydrogen and helium, with a methane and ammonia atmosphere.

Prior to 1989, very little else was known about Neptune. The Voyager 2 spacecraft, which conducted close-up flybys of Jupiter, Saturn and Uranus in 1979, 1981 and 1986, turned its cameras on Neptune in June 1989 and flew to within 3044 miles (4870 km) of the planet on 25 August. During this short time, mankind's knowledge of Neptune and its moons increased one hundred fold. For example, six moons and four rings were discovered, and the existence was confirmed of a magnetic field tilted 50 degrees from Neptune's rotational axis and offset 6000 miles (9600 km) from the planet's center. The temperature of Neptune was determined to be −353 degrees F (−178 degrees C).

Most notable visually was the discovery of the Great Dark Spot (GDS) in Neptune's atmosphere. Located at a mean latitude of 22 degrees south and with an overall length of 30 degrees longitude, it is very much

At right: Neptune, imaged by Voyager 2 on 23 August 1989—toward the top of the photo is the Great Dark Spot, a convection storm similar to Jupiter's Great Red Spot, with accompanying methane cirrus clouds; just below this, the speedy, bright cloud known as 'Scooter'; and below this, the Lesser Dark Spot, a convection storm with a permanent cloud bank over its center. See text, pages 102—104. Neptune, like Uranus, has a methane and ammonia atmosphere over a hydrogen and helium ocean.

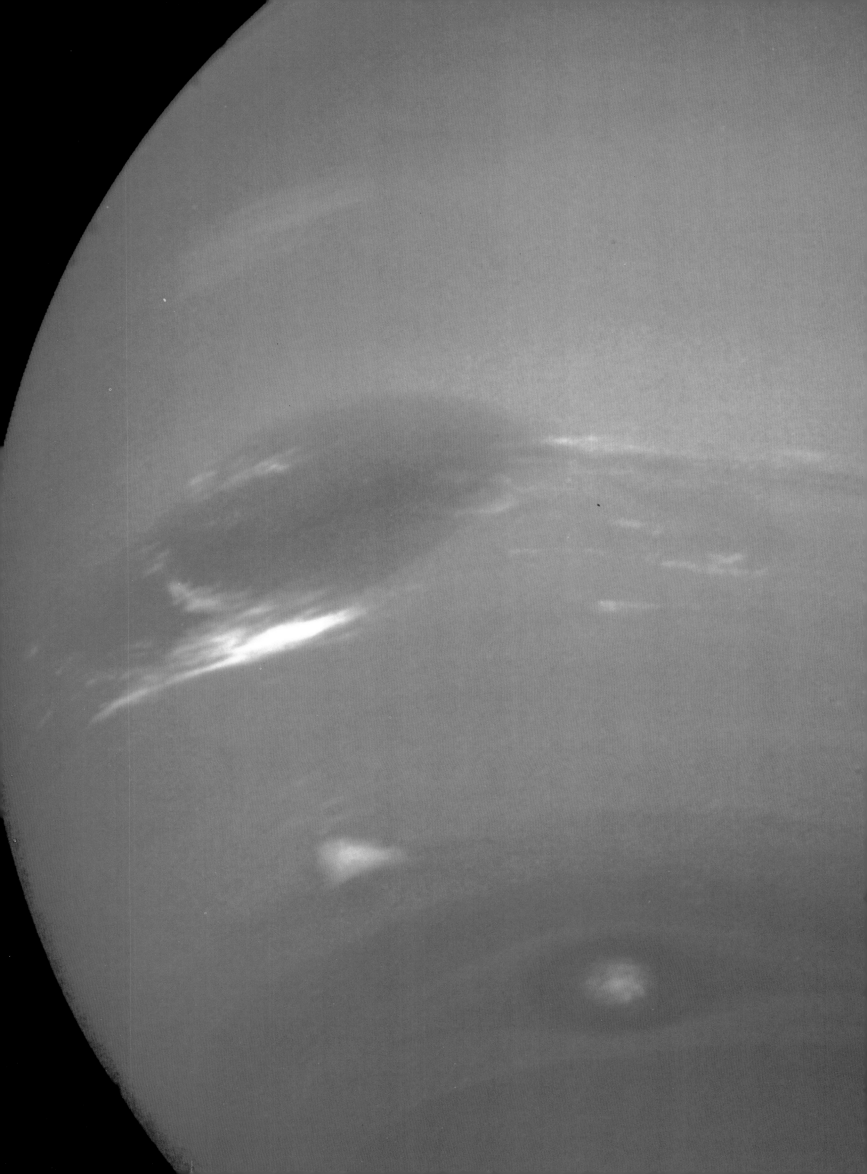

analogous to Jupiter's Great Red Spot (GRS), both in terms of location and its size relative to the planet. Like the GRS, the GDS is an elliptical, stormlike feature that rotates in a counterclockwise direction and is probably located *above* the surrounding cloud tops. The GDS is encircled by a constantly changing pattern of cirrus clouds and followed in its movement through Neptune's atmosphere by a string of much smaller elliptical storms. The cirrus clouds are composed of methane crystals and cling to the GDS the way cirrus cling to mountain tops on the Earth's islands. The Great Dark Spot completes a revolution of the planet in 18.3 hours, moving east to west.

To the south of the Great Dark Spot is another bright cloud which was observed by Voyager for the entire duration of the flyby and which was nicknamed 'scooter' because it moves at a much faster relative speed than the Great Dark Spot, causing it to overtake the latter every five days. South of this feature is the Lesser Dark Spot with a permanent cloud bank situated over its center.

Because Jupiter, Saturn and Uranus are all encircled by rings of rocky debris, it was long supposed of Neptune as well, although the planet is much too distant for these to be visible from Earth. It was not until two weeks prior to Voyager 2's closest encounter in August 1989 that the existence of these rings was verified. Because Neptune's rings are very

At top, above: A backlit Voyager 2 time exposure of the two main rings of Neptune, also showing an inner ring (about 25,000 miles [40,000 km] out from the center of Neptune) and a faint band that extends from 33,000 miles (52,800 km) to halfway between the two bright rings. *At right:* The Great Dark Spot, imaged by Voyager 2 during the spacecraft's transit over the face of Neptune, and 45 hours before its closest approach, 3000 miles (4800 km) — 'up and over' the planet's north pole.

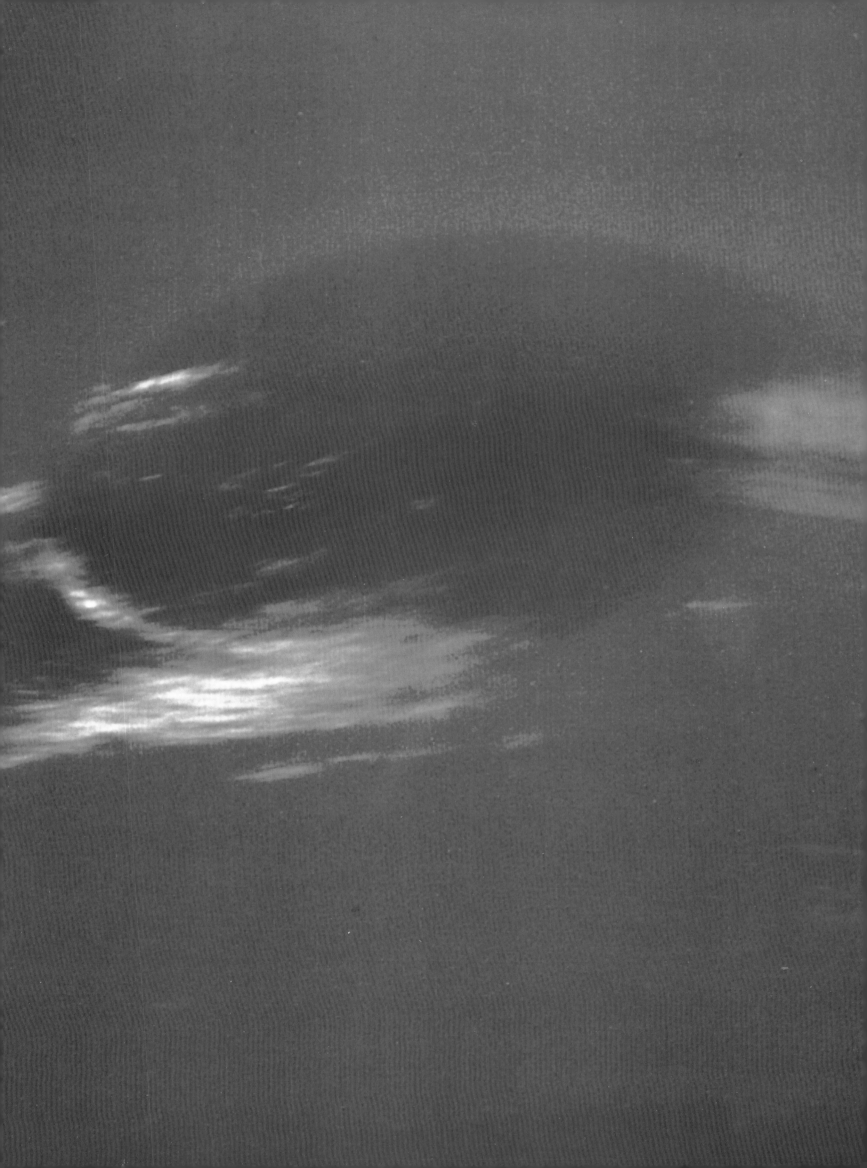

Above: A Voyager view approaching Neptune, showing the south pole, the
Lesser Dark Spot and the Great Dark Spot (which circles the planet in 18.3
hours), plus atmospheric banding. Compared to Earth, Neptune has 57 times
the volume and 17.2 times the mass; yet its density is just slightly more than
that of water—Earth has 4.8 times the density of this outermost gas giant. A
day on Neptune lasts 16.3 Earth hours, a year is 165 Earth years long and the

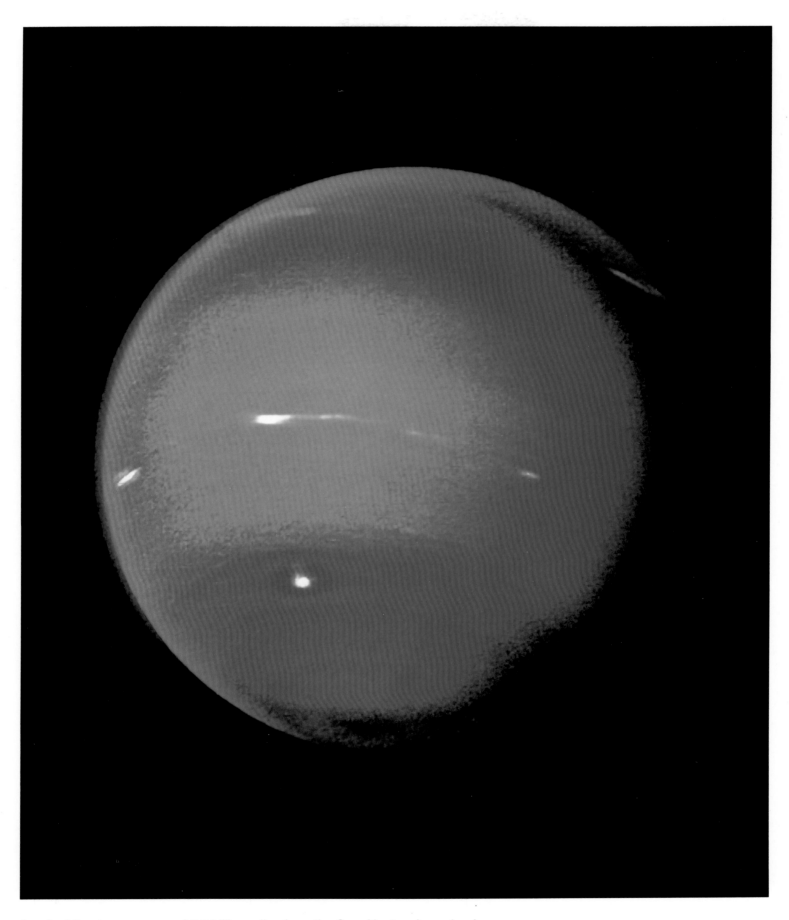

planet orbits at an average of 2.8 billion miles from the Sun. Neptune's angle of inclination is approximately 29 degrees (compared to 23 degrees for Earth), and its seasons last approximately 40 years.

Above: A Voyager south polar view leaving Neptune: false-color imaging through blue, green and methane-wavelength filters reveals concentrations of methane as bright patches.

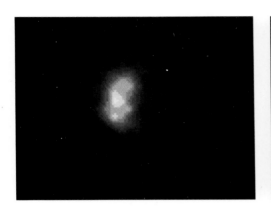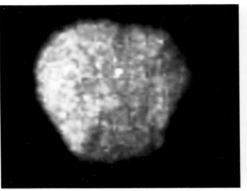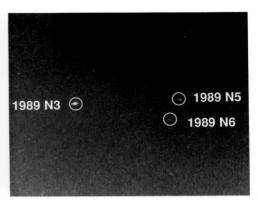

irregular, they appeared at first to be incomplete arcs. It is now known that there are four continuous, albeit thin, rings around the planet.

The two Neptunian moons confirmed prior to 1989 are among the most peculiar in the Solar System. Triton, discovered in 1846, less than a month after Neptune, is the largest of only two moons in the Solar System (the other is Saturn's Phoebe), to have a retrograde orbit around its mother planet; while Nereid, discovered more than a century after Triton, has the most elliptical orbit of any known moon in the Solar System. Voyager 2 discovered six additional moons, including the one known as 1989N1, which is actually larger than Nereid. Being darker, it is however, not visible from Earth. It is also the largest known nonsymmetrical body in the Solar System.

Triton, which revolves around Neptune in a direction opposite to the mother planet's rotation, was thought to have a diameter of 3700 miles (6000 km), but which is now known to be 1690 miles (2700 km). Triton is the second moon in the Solar System (after Saturn's Titan) that has been found to possess an atmosphere. It is a much thinner atmosphere than Titan's but is nevertheless clearly discernible in Voyager 2 photos of Triton's horizon. The atmosphere consists of methane chilled to −400 degrees F (−200 degrees C), making Triton the coldest object yet observed in the Solar System. Triton consists primarily of silicate rock, but there is also a great deal of water ice and frozen methane present.

Triton has a very pronounced division between that part of the moon which is experiencing summer and that part which is enduring winter. This phenomenon is not yet explained, nor is the fact that within the ten years prior to Voyager's flyby, Triton was twice as orange as it appeared both from Earth and from Voyager in August 1989. Evidence of recent volcanic activity on Triton has also been discovered, so that it is now possible to add it to the short list of volcanically active bodies which previously included only the Earth and Jupiter's moon Io.

 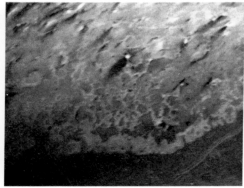 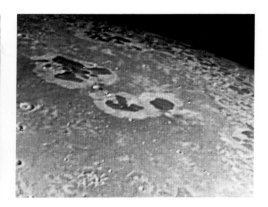

The Moons of Neptune

(NOTE: The numerically designated moons are listed in order of distance, but numbered in order of discovery.)

	Discovery	Diameter	Distance from Neptune
1989N6	Project Voyager, 1989	30 mi (50 km)	14,400 mi (23,000 km)
1989N5	Project Voyager, 1989	60 mi (90 km)	15,500 mi (25,000 km)
1989N3	Project Voyager, 1989	85 mi (135 km)	16,980 mi (27,150 km)
1989N4	Project Voyager, 1989	100 mi (160 km)	23,180 mi (37,100 km)
1989N2	Project Voyager, 1989	125 mi (200 km)	30,080 mi (48,100 km)
1989N1	Project Voyager, 1989	260 mi (420 km)	57,780 mi (92,500 km)
Triton	William Lassell, 1846	1690 mi (2700 km)	205,020 mi (329,880 km)
Nereid	Gerard Kuiper, 1949	300 mi (500 km)	3.4 million mi (5.5 million km)

Facing page, at top, left to right: Nereid, the outermost Neptunian moon, with an elliptical orbit between 86,982 and 5.9 million miles (139,171 and 9.4 million km) out; the asymmetrical moon 1989N1, from 57,780 miles (92,500 km); and the moons 1989N3, 1989N5 and 1989N6.

At top, above: A Voyager false-color image of Neptune's largest moon, Triton, with a retrograde orbit and an atmosphere. Triton is volcanically active, with one observed five-mile-high (eight km) eruption plume that drifted 90 miles (144 km) downwind. *At top center:* A Voyager false-color close-up of Triton: the long dark streaks could be windblown material and the smooth dark patches appear to be volcanic deposits. *At top, right:* A view taking in 600 miles (960 km) of Triton's surface. *Overleaf:* A Voyager mosaic of Triton's south pole. The light terrain may be polar ice.

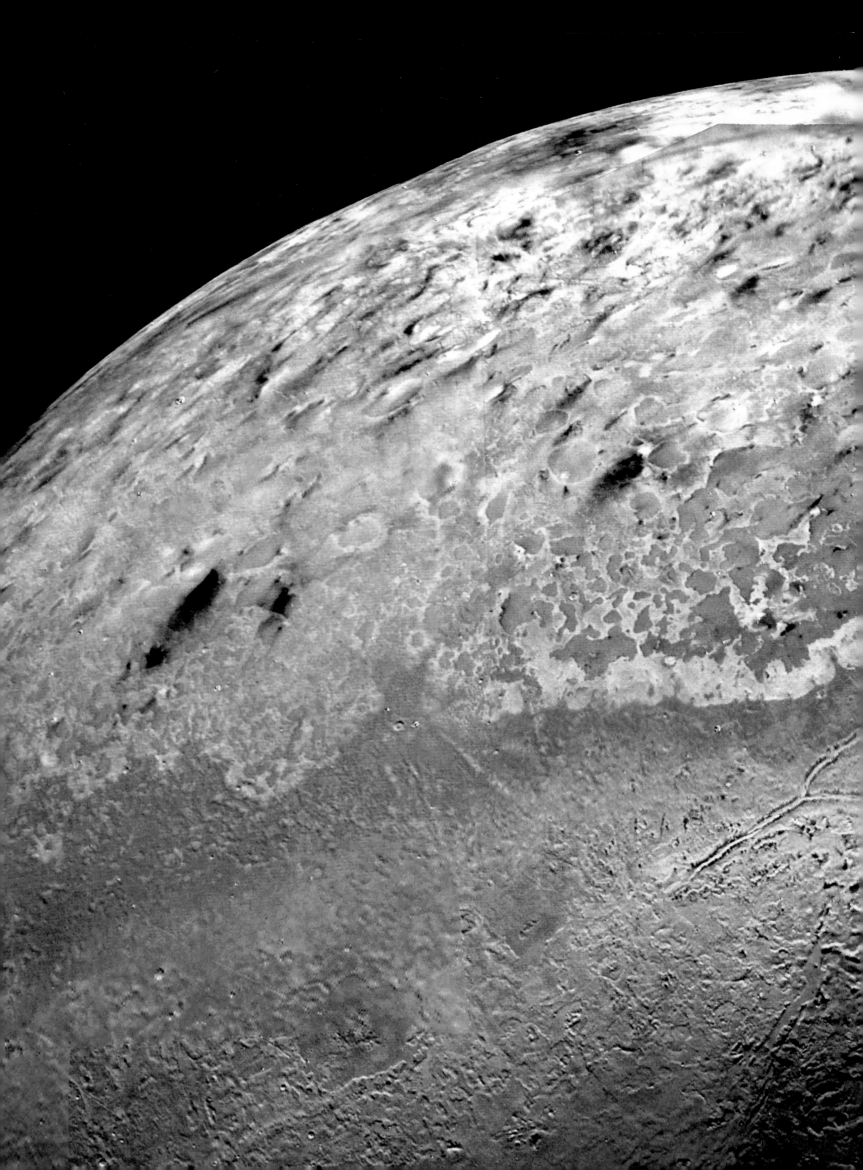

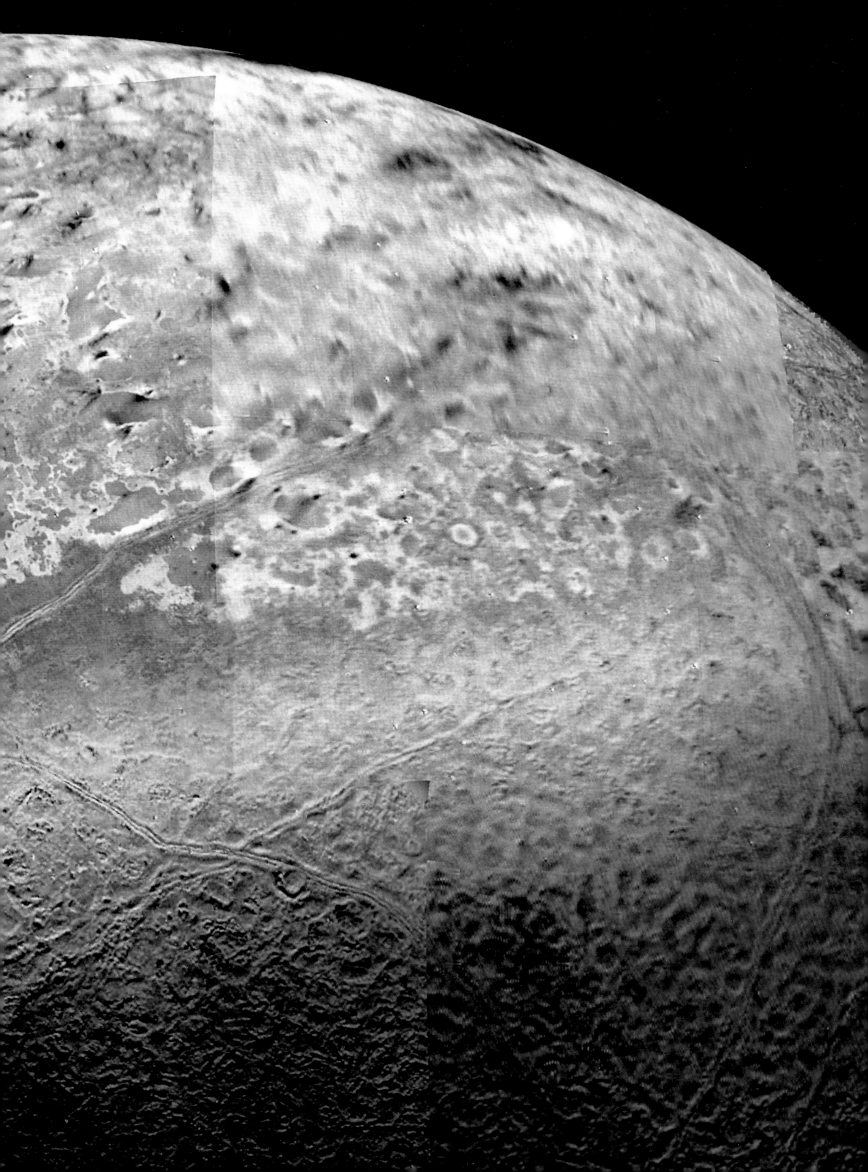

PLUTO AND BEYOND

In the decades that followed the discovery of Neptune, a great deal of astronomical effort went into attempts to locate additional planets. It seemed only logical that other planets existed beyond Neptune, and that finding another one was only a matter of time. However, it was a matter of a good deal *more* time than anyone could have predicted.

By the late nineteenth century astronomers charted slight anomalies in the revolutions of Uranus and Neptune, which were traceable to the gravitational effect of another body farther out in the Solar System. Around the turn of the century Percival Lowell embarked on a systematic search of the heavens, probing for what he called 'Planet X.' When Lowell died in 1916, others continued the search, including William Pickering of Harvard, who called the yet undiscovered object 'Planet O.' In 1915, and again in 1919, Pluto was actually photographed but not recognized because it was much fainter than it had been predicted to be. By this time, the organized search for Planet X was largely abandoned. In the meantime, Pickering altered his theory regarding the hypothetical location of Planet 'O,' and for the first time predicted that the perihelion of its orbit might actually bring it briefly closer to the Sun than Neptune. It was a radical idea that turned out to be accurate for Pluto.

In 1929 the Lowell Observatory at Flagstaff, Arizona, resumed the search begun by its founder, using a 13-inch telescope and a wide-field survey camera. This proved to be the right approach, and on 18 February 1930 the young astronomer Clyde Tombaugh identified a new planet in some photographs he had taken the previous month. The discovery was announced a month later on the 149th anniversary of the discovery of Uranus, and the new planet was called Pluto after the Roman god of the dead and the ruler of the underworld. The name was considered appropriate because of the planet's enormous distance from the Sun's warmth, and also because the first two letters were Percival Lowell's initials.

Pluto is just a speck in 'the great beyond,' just as the Solar System is a mere grain of sand on the shore of the Universal ocean. *At right:* The Pleiades, or 'Seven Sisters,' in the constellation Taurus. With a radius of 25 light years—147 trillion miles (235 trillion km)—this open star cluster is 410 light years distant—or 2.4 quadrillion miles (3.8 quadrillion km) from Earth. The Solar System, as far as Pluto, has a relatively small expanse of .0013 light years.

In the first years after it was discovered, physical data about Pluto was virtually impossible to obtain. In 1950, however, Gerard Kuiper at the Mount Palomar Observatory estimated its diameter at 3658 miles (5853 km), making it the second smallest planet in the Solar System. In 1965 it was observed in occultation with a 15th magnitude star, confirming that its diameter could not exceed 4200 miles (6700 km). Thus it was that the 3658 mile (5853 km) estimate held until the 1970s.

In 1976 methane ice was discovered to exist on Pluto's surface. Until then the planet's faintness had been attributed to its being composed of dark rock. Since ice would tend to reflect light moreso than dark rock, it would follow that if it *were* 3658 miles (5853 km) in diameter *and* covered with methane ice, it would be brighter than it actually is. Therefore, it was decided that Pluto was smaller than originally suspected, leading us to conclude that its diameter is less than the 2160 mile (3456 km) diameter of the Earth's Moon, and perhaps as small as 1375 miles (2200 km). This would make it the smallest of the nine planets and smaller than *seven* of the planetary moons.

As estimates of Pluto's size continue to be revised downward, it becomes less and less likely that it has the mass to exert the gravitational force on Neptune's orbit that was originally projected. If true, this would mean that Pluto is *not* 'Planet X,' and that 'Planet X' still exists and is yet to be discovered.

It has also been suggested that Pluto is perhaps the largest of a theorized belt of trans-Neptunian asteroids. However, that notion fails to take into account that Pluto is two and one-half times the diameter of Ceres, the largest known asteroid, and nearly seven times larger in diameter than the average of the 18 largest known asteroids.

Among the arguments that *can* be made for its not being a planet, or at least for its not being a 'normal' planet, are the peculiar aspects of its behavior. Its extremely elliptical orbit ranges from an aphelion of 4.6 million miles (7.4 million km) to a perihelion of 2.7 million miles (4.4 million km). The latter actually brings Pluto closer to the Sun than the perihelion of Neptune's much more circular orbit, as Pickering had believed. A second aspect of Pluto's behavior that sets it apart from other planets is its steep inclination to the elliptic plane. The orbits of all the planets are within two and one-half degrees of this same plane, except Mercury, which is inclined at seven degrees, and Pluto, which is inclined at an acute 17 degrees, making it very unusual among its peers.

A theory concerning the physical nature of Pluto holds that at one time

it was actually one of the moons of Neptune. It is further theorized that Pluto was somehow thrown out of its Neptunian orbit by some calamitous interaction with Neptune's moon Triton—perhaps even a collision. One of the Solar System's largest moons, Triton is larger than Pluto and, as such, might have had the gravitational force to slam a competing object out of Neptunian orbit if it ventured close enough. Both Triton and one of Neptune's other moons, Nereid, have unusual orbits that might possibly be relics of such a colossal event.

While its behavior partially defines it, and certainly sets it apart from other planets, less is known about Pluto's physical characteristics than is known about any other planet. Since no spacecraft will visit it in the twentieth century, we are left with only educated guesses about Pluto.

We know that it is extremely cold, with noontime summer temperatures rarely creeping above −350 degrees F (−175 degrees C). Its rocky surface is known to also contain methane, probably in the form of ice or frost. Water ice may also be present, though this is not likely, and Pluto's mass suggests a rocky core. Pluto has generally been thought to have no atmosphere because its relatively small mass wouldn't give it

At top, above: Clyde Tombaugh's Pluto discovery photos, on 18 February 1930. The arrows track Pluto on its migration between exposures (23 and 29 January 1930). *Overleaf:* An artist's concept of Pluto and its moon Charon, whose 153-hour orbit matches Pluto's rotational period: the same side of Charon faces the same side of Pluto always. Charon may be as large as 744 miles (1190 km)—roughly half Pluto's diameter.

sufficient gravity to retain an atmosphere, and it is too cold for even such a substance as methane to easily exist in its gaseous state. However, Scott Sawyer of the University of Texas has discovered what may be a tenuous methane vapor atmosphere on Pluto.

For a traveler from Earth, Pluto would be a distant and lonely place worthy of the god of the netherworld. It is a frigid sphere where the Sun is scarcely brighter than the most luminous star as viewed from Earth, and from which radio communication transmitted at the speed of light would take five hours to reach the Earth. Indeed, the travel time to Pluto, using current state of the art technology, would be 10 Earth years or more.

For Pluto itself, a year—the time required for it to complete a revolution around the Sun—is equal to 248 Earth years, although its day—or rotational period—is estimated at only 6.3 times that of Earth.

The discovery of the Plutonian moon Charon came about indirectly in 1978. While James Christy at the US Naval Observatory in Flagstaff, Arizona was attempting to measure Pluto's size, he thought he'd noticed that it was not spherical. Further observations led him to the conclusion that the elongation he had observed was due to the presence of a satellite very close to Pluto. Further calculations indicated that this newly discovered body was as close as 10,563 miles (16,900 km) from Pluto.

Like Pluto itself, the environment beyond it remains a mystery in terms of planetary discovery. If Pluto is not Planet X, and Percival Lowell's original theory still holds, then another object the size of Mars—or larger—awaits us in the dim distance beyond the five billion mile (eight billion km) mark. With Pluto still such a mystery, it is hard to speculate about what such a world might be like.

We don't really know how many planets might populate this distant netherworld at the limits of the gravitational effect of our Sun. We now know of nine planets, but past experience has taught us to be hesitant about stating that there are *only* nine. Furthermore, we know that our Sun is a star like billions of other stars in the universe, so it is entirely possible that we will one day encounter other solar systems—perhaps a vast number of other solar systems—in orbit around other suns.

At right: The final Voyager 2 image of Neptune's moon, Triton. Voyager won't be visiting Pluto, which orbits an average of 3.6 billion miles (5.9 billion km) from the Sun. Some observers consider Triton to be probably very similar to the planet Pluto—a notion that would seem to mollify the need for close observation. It is mere placebo, as we won't see Pluto close-up in our lifetime, and know almost nothing of Pluto.

Index

Final page: A Voyager post-encounter view of the south pole of Neptune. Voyager flew toward, over and away from the planet in a broad bell curve, then sped off into the far outer Solar System. By 1996, Voyager will have reached the magnetosphere-like boundary between the solar wind and the onrushing interstellar medium—the *helio-pause*—a place and a set of circumstances not currently understood. This *may* then be the outer limit of the Solar System—a definition we hesitate to use.

THE
SOLAR SYSTEM

The diagrams below show the known objects that are within 40 Astronomical Units (3.7 billion miles) of the Sun. Among known objects only Pluto at aphelion is farther from the Sun. The top (and largest) section of the diagram shows the relative distances of the Solar System's moons from their respective planets on a *15 million mile scale*. Most of them, however, are nearer than a million miles.

The center diagram shows the planets themselves on a *40 Astronomical Unit scale,* while the bottom

The Moons of the Solar System

Scale				
15 Million Miles		The Moons of **Jupiter:** *Sinope *Pasiphae		
14 Million Miles		*Carme		
13 Million Miles				
12 Million Miles		*Ananke		
11 Million Miles				
10 Million Miles				
9 Million Miles			The Moons of **Saturn:** * Phoebe	
8 Million Miles		*Elara *Lysithea *Himalia		
7 Million Miles		*Leda		
6 Million Miles				
5 Million Miles				
4 Million Miles				The Moons of **Uranus:**
3 Million Miles				Oberon / Titania / Umbriel / Ariel / Miranda
2 Million Miles			* Iapetus	Puck / Belinda / Cressida / Portia / Rosalind
		* Callisto		
The Moon of the **Earth:**	The Moons of **Mars:** Deimos Phobos	Ganymede / Amalthea / Europa / Adrastea / Io / Metis / Thebe	Janus / Telesto / Rhea / Epimetheus * Tethys / Helene / Pandora / Enceladus / Dione Co-orbital / Prometheus / Mimas Co-orbital / Dione / Hyperion / Atlas / Mimas / Calypso / Titan	Desdemona / Juliet / Bianca / Ophelia / Cordelia

The Planets of the Solar System

Sun — 1 — Mercury — Venus — Earth — 2 — Mars — 3 — 4 — 5 — Jupiter — 6 — 7 — 8 — 9 — Saturn — 10 — 11 — 12 — 13 — 14 — 15 — 16 — 17 — 18 — 19 — Uranus — 20

The Asteroid Belts

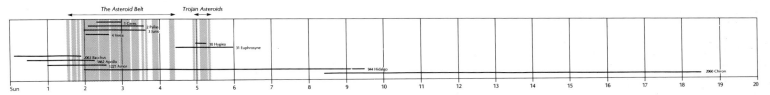

The Asteroid Belt — Trojan Asteroids

1 Ceres / 2 Pallas / 3 Juno / 4 Vesta / 10 Hygiea / 31 Euphrosyne / 2063 Bacchus / 1862 Apollo / 1221 Amor / 944 Hidalgo / 2060 Chiron

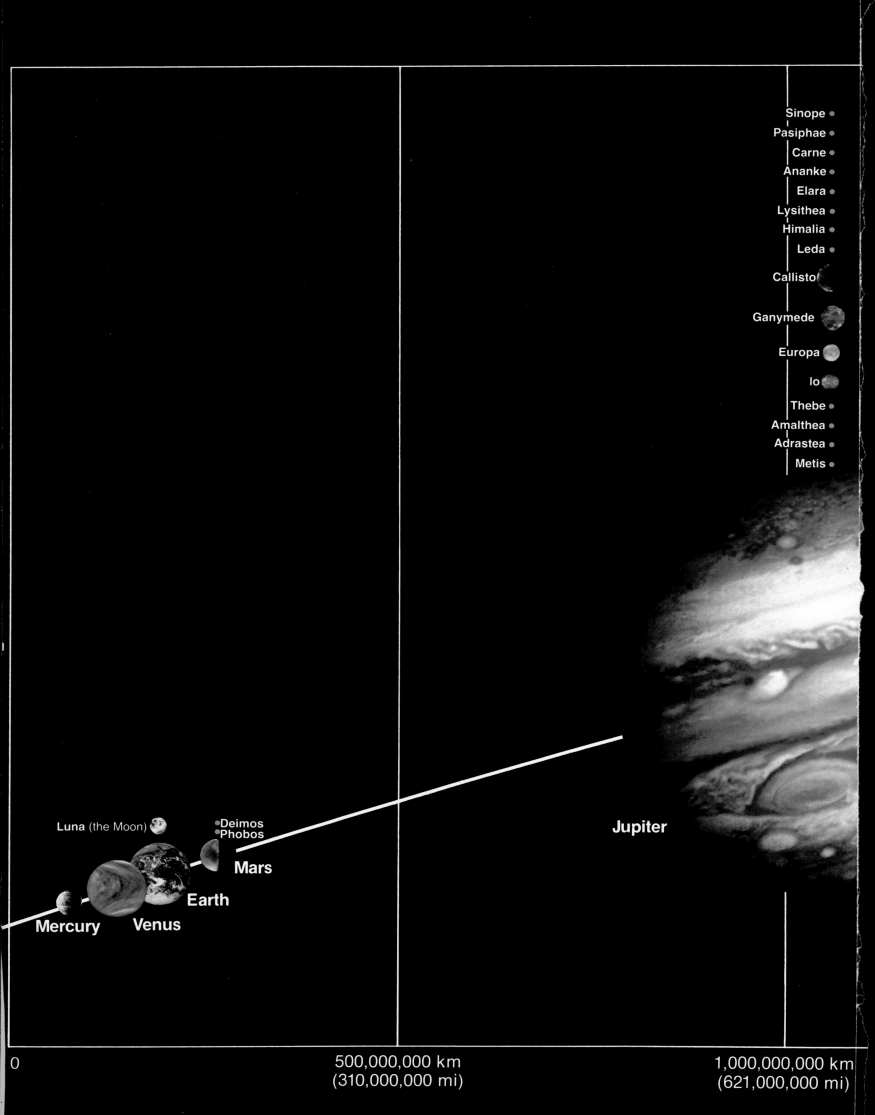

Sinope •
Pasiphae •
Carne •
Ananke •
Elara •
Lysithea •
Himalia •
Leda •
Callisto
Ganymede
Europa
Io
Thebe •
Amalthea •
Adrastea •
Metis •

Jupiter

Luna (the Moon)

•Deimos
•Phobos

Mars

Earth

Mercury   Venus

0

500,000,000 km
(310,000,000 mi)

1,000,000,000 km
(621,000,000 mi)

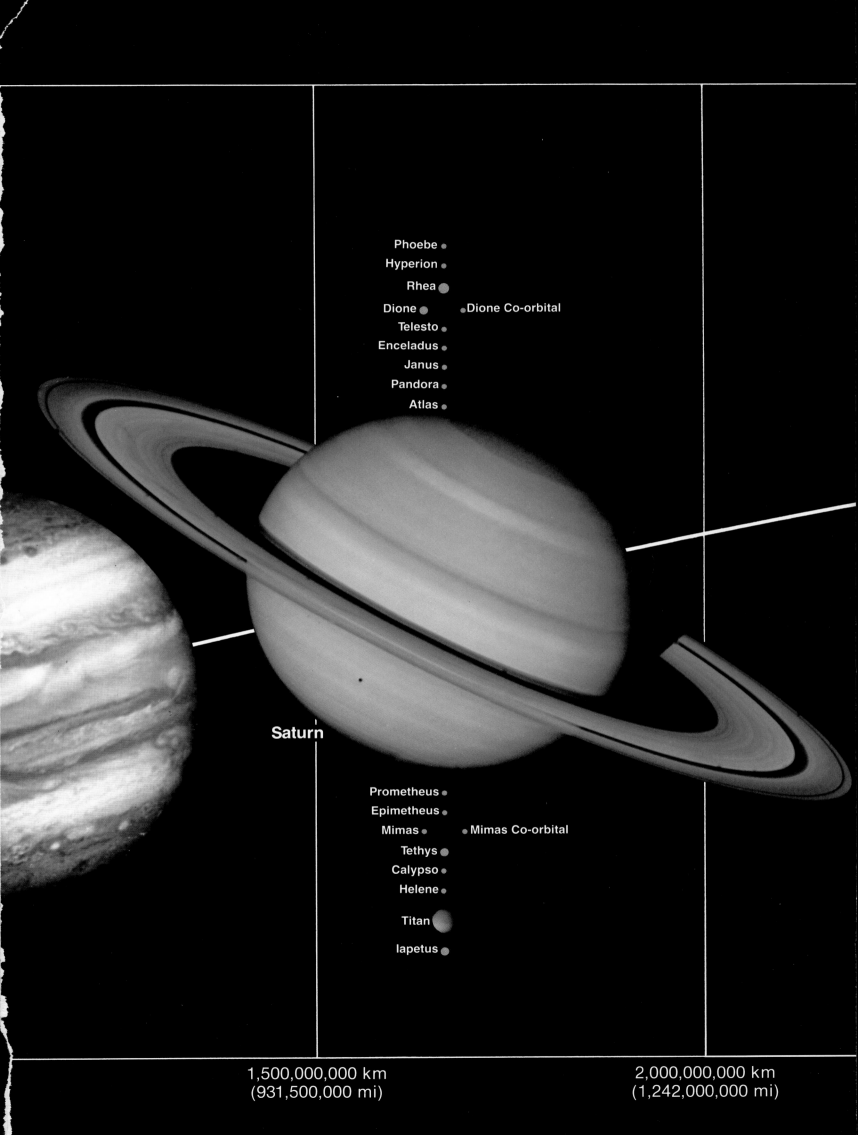

Phoebe ●
Hyperion ●
Rhea ●
Dione ●　　●Dione Co-orbital
Telesto ●
Enceladus ●
Janus ●
Pandora ●
Atlas ●

Saturn

Prometheus ●
Epimetheus ●
Mimas ●　　● Mimas Co-orbital
Tethys ●
Calypso ●
Helene ●

Titan ●

Iapetus ●

1,500,000,000 km
(931,500,000 mi)

2,000,000,000 km
(1,242,000,000 mi)

diagram shows a selection of asteroids on the same scale. The shaded area on the bottom two diagrams is the Asteroid Belt.

The asteroids are indicated by lines rather than dots because they have highly elliptical orbits and hence a big difference between their respective aphelions and perihelions. In the case of 1 Ceres, the largest of the asteroids, this difference is roughly one-half AU, but 2060 Chiron swings wildly in an orbit that varies by 10 AU, or nearly a billion miles!

(Million-mile vertical scale)

(Astronomical Unit horizontal scale)

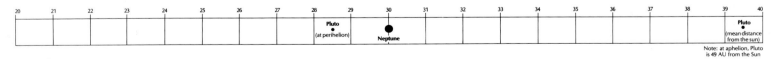

Note: at aphelion, Pluto is 49 AU from the Sun

(Astronomical Unit horizontal scale)

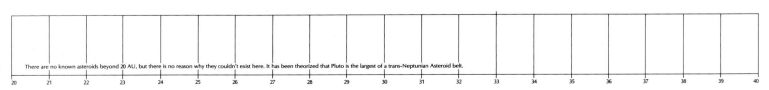

There are no known asteroids beyond 20 AU, but there is no reason why they couldn't exist here. It has been theorized that Pluto is the largest of a trans-Neptunian Asteroid belt.

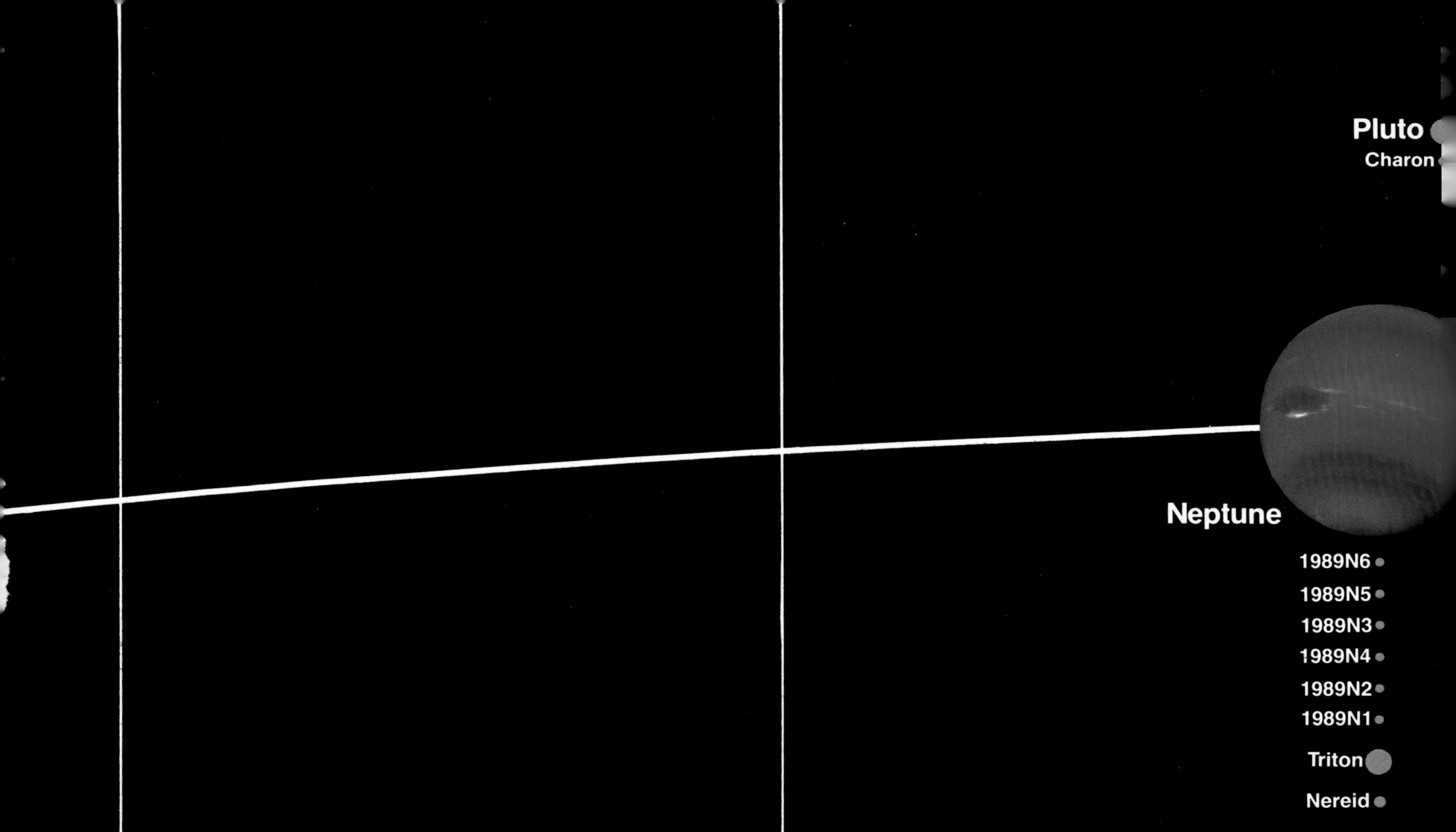

Pluto

Charon

Neptune

1989N6
1989N5
1989N3
1989N4
1989N2
1989N1

Triton

Nereid

Notes: All objects are shown to scale with the exceptions of the planets Jupiter, Saturn, Uranus and Neptune which are shown at 60 percent of their actual sizes relative to their moons and the remaining planets.

The planets are shown at their approximate mean distances from the sun. This scale does not correspond with the object size scale.

Moons are shown in order of their orbital distances. These distances are not shown to scale.

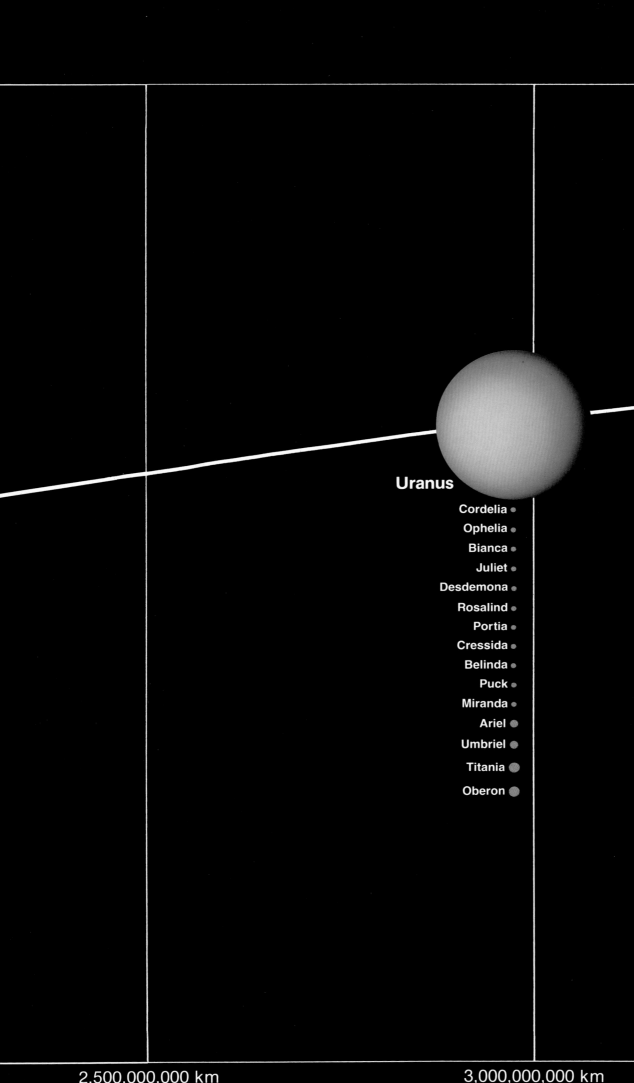

Uranus

Cordelia ●
Ophelia ●
Bianca ●
Juliet ●
Desdemona ●
Rosalind ●
Portia ●
Cressida ●
Belinda ●
Puck ●
Miranda ●
Ariel ●
Umbriel ●
Titania ●
Oberon ●

2,500,000,000 km   3,000,000,000 km